COVER CAPTION:

Map of annular and total solar eclipses of 2010.

The NASA STI Program Office ... in Profile

Since its founding, NASA has been dedicated to the advancement of aeronautics and space science. The NASA Scientific and Technical Information (STI) Program Office plays a key part in helping NASA maintain this important role.

The NASA STI Program Office is operated by Langley Research Center, the lead center for NASA's scientific and technical information. The NASA STI Program Office provides access to the NASA STI Database, the largest collection of aeronautical and space science STI in the world. The Program Office is also NASA's institutional mechanism for disseminating the results of its research and development activities. These results are published by NASA in the NASA STI Report Series, which includes the following report types:

- TECHNICAL PUBLICATION. Reports of completed research or a major significant phase of research that present the results of NASA programs and include extensive data or theoretical analysis. Includes compilations of significant scientific and technical data and information deemed to be of continuing reference value. NASA's counterpart of peer-reviewed formal professional papers but has less stringent limitations on manuscript length and extent of graphic presentations.

- TECHNICAL MEMORANDUM. Scientific and technical findings that are preliminary or of specialized interest, e.g., quick release reports, working papers, and bibliographies that contain minimal annotation. Does not contain extensive analysis.

- CONTRACTOR REPORT. Scientific and technical findings by NASA-sponsored contractors and grantees.

- CONFERENCE PUBLICATION. Collected papers from scientific and technical conferences, symposia, seminars, or other meetings sponsored or cosponsored by NASA.

- SPECIAL PUBLICATION. Scientific, technical, or historical information from NASA programs, projects, and mission, often concerned with subjects having substantial public interest.

- TECHNICAL TRANSLATION. English-language translations of foreign scientific and technical material pertinent to NASA's mission.

Specialized services that complement the STI Program Office's diverse offerings include creating custom thesauri, building customized databases, organizing and publishing research results . . . even providing videos.

For more information about the NASA STI Program Office, see the following:

- Access the NASA STI Program Home Page at http://www.sti.nasa.gov/STI-homepage.html

- E-mail your question via the Internet to help@sti.nasa.gov

- Fax your question to the NASA Access Help Desk at (301) 621-0134

- Telephone the NASA Access Help Desk at (301) 621-0390

- Write to:
  NASA Access Help Desk
  NASA Center for AeroSpace Information
  7115 Standard Drive
  Hanover, MD 21076–1320

NASA/TP—2008–214171

Annular and Total Solar Eclipses of 2010

F. Espenak
NASA Goddard Space Flight Center, Greenbelt, Maryland

J. Anderson
Royal Astronomical Society of Canada, Winnipeg, Manitoba

National Aeronautics and
Space Administration

Goddard Space Flight Center
Greenbelt, Maryland 20771

November 2008

Available from:

NASA Center for AeroSpace Information	National Technical Information Service
7115 Standard Drive	5285 Port Royal Road
Hanover, MD 21076-1320	Springfield, VA 22161

Reader's Note

While most NASA eclipse bulletins cover a single eclipse, this publication presents predictions for two solar eclipses during 2010. This has required a different organization of the material into the following sections.

Section 1—Eclipse Predictions: The section consists of a general discussion about the eclipse path maps, Besselian elements, shadow contacts, eclipse path tables, local circumstances tables, and the lunar limb profile.

Section 2—Annular Solar Eclipse of 2010 Jan 15: The section covers predictions and weather prospects for the annular eclipse.

Section 3—Total Solar Eclipse of 2010 Jul 11: The section covers predictions and weather prospects for the total eclipse.

Section 4—Observing Eclipses: The section provides information on eye safety, solar filters, eclipse photography, and making contact timings from the path limits.

Section 5—Eclipse Resources: The final section contains a number of resources including information on the IAU Working Group on Eclipses, the Solar Eclipse Mailing List, the NASA eclipse bulletins on the Internet, Web sites for the two 2010 eclipses, and a summary identifying the algorithms, ephemerides, and parameters used in the eclipse predictions.

Bold headers at the top of every page identify the section number and title for easy reference. This should help the reader to quickly navigate to any section.

Table of Contents

PREFACE vii

1. ECLIPSE PREDICTIONS 1
    1.1 Introduction 1
    1.2 Orthographic Projection Maps 1
    1.3 Equidistant Conic Projection Maps of the Eclipse Paths 2
    1.4 Detailed Maps of the Eclipse Paths 2
    1.5 Elements, Shadow Contacts, and Eclipse Path Tables 2
    1.6 Local Circumstances Tables 3
    1.7 Mean Lunar Radius 4
    1.8 Lunar Limb Profile 5

2. ANNULAR SOLAR ECLIPSE OF 2010 JAN 15 7
    2.1 Introduction 7
    2.2 Antumbral Path and Visibility 7
    2.3 Maps of the Annular Eclipse Path 8
    2.4 Annular Eclipse Elements and Path Tables 8
    2.5 Annular Eclipse Local Circumstances Tables 8
    2.6 Annular Eclipse Lunar Limb Profile 8
    2.7 Weather Prospects for the Annular Eclipse 9

3. TOTAL SOLAR ECLIPSE OF 2010 JUL 11 43
    3.1 Introduction 43
    3.2 Umbral Path and Visibility 43
    3.3 Maps of the Total Eclipse Path 43
    3.4 Total Eclipse Elements and Path Tables 44
    3.5 Total Eclipse Local Circumstances Tables 44
    3.6 Total Eclipse Lunar Limb Profile 44
    3.7 Weather Prospects for the Total Eclipse 44

4. OBSERVING ECLIPSES 65
    4.1 Eye Safety and Solar Eclipses 65
    4.2 Sources for Solar Filters 67
    4.3 Eclipse Photography 67
    4.4 Contact Timings from the Path Limits 68
    4.5 Plotting Eclipse Paths on Maps 69
    4.6 Eclipse Paths on Google Maps 69

5. ECLIPSE RESOURCES 73
    5.1 IAU Working Group on Eclipses 73
    5.2 IAU Solar Eclipse Education Committee 73
    5.3 Solar Eclipse Mailing List 73
    5.4 NASA Eclipse Bulletins on the Internet 73
    5.5 Future Eclipse Paths on the Internet 73
    5.6 NASA Web Site for 2010 Solar Eclipses 74
    5.7 Predictions for Eclipse Experiments 74
    5.8 Algorithms, Ephemerides, and Parameters 74

AUTHOR'S NOTE 74
ACRONYMS and UNITS 75
REFERENCES
    Bibliography 75
    Further Reading on Eclipses 76
    Further Reading on Eye Safety 77
    Further Reading on Meteorology 77

Preface

This work is the thirteenth in a series of NASA publications containing detailed predictions, maps, and meteorological data for future total and annular solar eclipses of interest. Published as part of NASA's Technical Publication (TP) series, the eclipse bulletins are prepared in cooperation with the Working Group on Eclipses of the International Astronomical Union and are provided as a public service to both the professional and lay communities, including educators and the media. In order to allow a reasonable lead time for planning purposes, eclipse bulletins are published 12 to 24 months before each event.

Single copies of the bulletins are available at no cost by sending a 9 × 12 inch self-addressed stamped envelope with postage for 12 oz. (340 g). Detailed instructions and an order form can be found at the back of this publication.

The 2010 bulletin uses the World Data Bank II (WDBII) mapping database for the path figures. WDBII outline files were digitized from navigational charts to a scale of approximately 1:3,000,000. The database is available through the *Global Relief Data CD-ROM* from the National Geophysical Data Center. The highest detail eclipse maps are constructed from the Digital Chart of the World (DCW), a digital database of the world developed by the U.S. Defense Mapping Agency (DMA). The primary sources of information for the geographic database are the Operational Navigation Charts (ONC) and the Jet Navigation Charts (JNC). The eclipse path and DCW maps are plotted at a scale of 1:3,000,000 to 1:6,000,000 in order to show roads, cities, and villages, and lakes and rivers, making them suitable for eclipse expedition planning. Place names are from the World Gazetteer at <http://www.world-gazetteer.com/>.

The geographic coordinates database includes over 90,000 cities and locations. This permits the identification of many more cities within the umbral path and their subsequent inclusion in the local circumstances tables. Many of these locations are plotted in the path figures when the scale allows. The source of these coordinates is Rand McNally's *The New International Atlas*. A subset of these coordinates is available in digital form, which has been augmented with population data.

The bulletins have undergone a great deal of change since their inception in 1993. The expansion of the mapping and geographic coordinates databases have improved the coverage and level of detail. This renders them suitable for the accuracy required by scientific eclipse expeditions. Some of these changes are the direct result of suggestions from the end user. Readers are encouraged to share comments and suggestions on how to improve the content and layout in subsequent editions. Although every effort is made to ensure that the bulletins are as accurate as possible, an error occasionally slips by. We would appreciate your assistance in reporting all errors, regardless of their magnitude.

We thank Dr. B. Ralph Chou for a comprehensive discussion on solar eclipse eye safety (Sect. 4.1). Dr. Chou is Professor of Optometry at the University of Waterloo with over 30 years of eclipse observing experience. As a leading authority on the subject, Dr. Chou's contribution should help dispel much of the fear and misinformation about safe eclipse viewing.

The **NASA Eclipse Web Site** provides general information on every solar and lunar eclipse occurring during the period 1901 through 2100. An online catalog also lists the date and basic characteristics of every solar and lunar eclipse from 2000 BCE through 3000 CE. The *World Atlas of Solar Eclipses* provides maps for every central solar eclipse path over the same five-millennium period. The URL of the **NASA Eclipse Web Site** is <http://eclipse.gsfc.nasa.gov/>.

In addition to the synoptic data provided by the Web site above, special Web sites have been prepared for the annular and total solar eclipses of 2010. The URL of the annular eclipse Web site is <http://eclipse.gsfc.nasa.gov/SEmono/ASE2010/ASE2010.html>. The URL of the total eclipse Web site is <http://eclipse.gsfc.nasa.gov/SEmono/TSE2010/TSE2010.html>. These Web sites include supplemental predictions, figures, and maps, which are not included in the present publication.

Because the eclipse bulletins have size limits, they cannot include everything needed by every scientific investigation. Some investigators may require exact contact times, which include lunar limb effects, or times for a specific observing site not listed in the bulletin. Other investigations may need customized predictions for an aerial rendezvous, or near the path limits for grazing eclipse experiments. We would like to assist such investigations by offering to calculate additional predictions for any professionals or large groups of amateurs. Please contact Fred Espenak with complete details and eclipse prediction requirements.

We would like to acknowledge the valued contributions of a number of individuals who were essential to the success of this publication. The format and content of the NASA eclipse bulletins has drawn heavily from over 40 years of eclipse *Circulars* published by the U.S. Naval Observatory. We owe a debt of gratitude to past and present staff of that institution who performed this service for so many years. The numerous publications and algorithms of Dr. Jean Meeus have served to inspire a life-long interest in eclipse prediction. Prof. Jay M. Pasachoff reviewed the manuscript and offered many helpful suggestions. As Chair of the International Astronomical Union's (IAU) Working Group on Eclipses, Prof. Pasachoff maintains a general Web site at

Preface

<http://www.eclipses.info> that links to many eclipse related Web sites. Dr. David Dunham and Mr. Paul Maley reviewed and updated the information about eclipse contact timings. Ms. Elaine Firestone (Goddard Publications Senior Technical Editor) meticulously reviewed the manuscript. She was responsible for the editing, two-column page layout, and for ensuring that the bulletin conforms to NASA publication standards.

Permission is freely granted to reproduce any portion of this publication, including data, figures, maps, tables, and text. All uses and/or publication of this material should be accompanied by an appropriate acknowledgment (e.g., "Reprinted from NASA's *Annular and Total Solar Eclipses of 2010*, Espenak and Anderson 2008"). We would appreciate receiving a copy of any publications where this material appears.

The names and spellings of countries, cities, and other geopolitical regions are for identification purposes only. They are not authoritative, nor do they imply any official recognition in status by the United States Government. Corrections to names, geographic coordinates, and elevations are actively solicited in order to update the database for future bulletins. All calculations, diagrams, and opinions are those of the authors and they assume full responsibility for their accuracy.

Fred Espenak
NASA Goddard Space Flight Center
Planetary Systems Laboratory, Code 693
Greenbelt, MD 20771
USA

E-mail: fred.espenak@nasa.gov

Jay Anderson
Royal Astronomical Society of Canada
189 Kingsway Ave.
Winnipeg, MB
CANADA R3M 0G4

E-mail: jander@cc.umanitoba.ca

NASA Solar Eclipse Bulletins

NASA Eclipse Bulletin	RP #	Publication Date
Annular Solar Eclipse of 1994 May 10	1301	April 1993
Total Solar Eclipse of 1994 November 3	1318	October 1993
Total Solar Eclipse of 1995 October 24	1344	July 1994
Total Solar Eclipse of 1997 March 9	1369	July 1995
Total Solar Eclipse of 1998 February 26	1383	April 1996
Total Solar Eclipse of 1999 August 11	1398	March 1997

NASA Eclipse Bulletin	TP #	Publication Date
Total Solar Eclipse of 2001 June 21	1999-209484	November 1999
Total Solar Eclipse of 2002 December 04	2001-209990	October 2001
Annular and Total Solar Eclipses of 2003	2002-211618	October 2002
Total Solar Eclipse of 2006 March 29	2004-212762	November 2004
Total Solar Eclipse of 2008 August 01	2007-214149	March 2007
Total Solar Eclipse of 2009 July 22	2008-214169	March 2008
Annular and Total Solar Eclipses of 2010	2008-214171	November 2008

1. Eclipse Predictions

1.1 Introduction

During 2010, there are two major eclipses of the Sun. The first is an annular eclipse on January 15 (Section 2) and the second is a total eclipse on July 11 (Section 3). This section is a general description of the tables, maps, and figures appearing in the later sections for each eclipse.

For simplicity, the term "umbral" will be used when referring to either the "umbral" (total eclipse) or "antumbral" (annular eclipse) shadow or path.

1.2 Orthographic Projection Maps

Figures 2.1 and 3.1 feature an orthographic projection map of Earth (adapted from Espenak 1987) for the annular (Jan 15) and total (Jul 11) eclipses, respectively. Each map shows the path of penumbral (partial eclipse) and umbral shadows (annular or total eclipse). The daylight terminator is plotted for the instant of greatest eclipse with north at the top. The map is centered over the point of greatest eclipse and is indicated with an asterisk symbol. The subsolar point (Sun in zenith) at that instant is also shown.

The limits of the Moon's penumbral shadow define the region of visibility of the partial eclipse. This saddle-shaped region often covers more than half of Earth's daylight hemisphere and consists of several distinct zones or limits. At the northern and/or southern boundaries lie the limits of the penumbra's path. Partial eclipses have only one of these limits, as do umbral eclipses when the shadow axis falls no closer than about 0.45 radii from Earth's center. Great loops at the western and eastern extremes of the penumbra's path identify the areas where the eclipse begins and ends at sunrise and sunset, respectively.

In the case of the 2010 annular eclipse (Figure 2.1), the penumbra has both a northern and southern limit, so that the rising and setting curves form two separate, closed loops. In comparison, the penumbral shadow of the 2010 total eclipse (Figures 3.1) has no southern limit; its southern edge falls off Earth so the southern penumbral path is bounded by Earth's terminator.

Bisecting the "eclipse begins and ends at sunrise and sunset" loops is the curve of maximum eclipse at sunrise (western loop) and sunset (eastern loop). The exterior tangency points *P1* and *P4* mark the coordinates where the penumbral shadow first contacts (partial eclipse begins) and last contacts (partial eclipse ends) Earth's surface. The path of the umbral shadow bisects the penumbral path from west to east.

A curve of maximum eclipse is the locus of all points where the eclipse is at maximum at a given time. They are plotted at each half hour in Universal Time, and generally run in a north-south direction. The outline of the umbral shadow is plotted every 15 min in Universal Time. Curves of constant eclipse magnitude[1] delineate the locus of all points where the magnitude at maximum eclipse is constant. These curves run exclusively between the curves of maximum eclipse at sunrise and sunset. Furthermore, they are quasi-parallel to the southern penumbral limit. This limit may be thought of as a curve of constant magnitude of 0.0, while the adjacent curves are for magnitudes of 0.2, 0.4, 0.6, and 0.8. The northern and southern limits of the path of total eclipse are curves of constant magnitude of 1.0.

At the top of Figures 2.1 and 3.1, the Universal Time of geocentric conjunction in ecliptic coordinates between the Moon and Sun is given (i.e., instant of New Moon) followed by the instant of greatest eclipse. The eclipse magnitude is given for greatest eclipse. It is equivalent to the topocentric ratio of diameters of the Moon and Sun. Gamma is the minimum distance of the Moon's shadow axis from Earth's center in units of equatorial Earth radii. Finally, the Saros series number of the eclipse is given along with its relative sequence in the series.

In the upper left and right corners are the geocentric coordinates of the Sun and Moon, respectively, at the instant of greatest eclipse. They are:

R.A.—Right Ascension
Dec.—Declination
S.D.—Apparent Semi-Diameter
H.P.—Horizontal Parallax

To the lower left are the exterior/interior contact times of the Moon's penumbral shadow with Earth, which are defined:

P1—Instant of first exterior tangency of Penumbra with Earth's limb. (Partial Eclipse Begins)
P2—Instant of first interior tangency of Penumbra with Earth's limb.
P3—Instant of last interior tangency of Penumbra with Earth's limb.
P4—Instant of last exterior tangency of Penumbra with Earth's limb. (Partial Eclipse Ends)

Not all eclipses have P2 and P3 penumbral contacts. They are only present in cases where the penumbral shadow falls completely within Earth's disk. For instance, the 2010 annular eclipse (Jan 15) has all four penumbral contacts, but the total eclipse (Jul 11) only has P1 and P4. The lower right corner lists exterior/interior contact times of the Moon's umbral shadow with Earth's limb which are defined as follows:

U1—Instant of first exterior tangency of Umbra with Earth's limb. (Umbral [Total/Annular] Eclipse Begins)
U2—Instant of first interior tangency of Umbra with Earth's limb.
U3—Instant of last interior tangency of Umbra with Earth's limb.
U4—Instant of last exterior tangency of Umbra with Earth's limb. (Umbral [Total/Annular] Eclipse Ends)

---

1. Eclipse magnitude is defined as the fraction of the Sun's diameter occulted by the Moon. It is strictly a ratio of diameters and should not be confused with eclipse obscuration, which is a measure of the Sun's surface *area* occulted by the Moon. Eclipse magnitude is usually expressed as a decimal fraction (e.g., 0.50 for 50%).

At bottom center are the geographic coordinates of the position of greatest eclipse, along with the local circumstances at that location (i.e., Sun altitude, Sun azimuth, path width, and duration of totality/annularity). At bottom left are a list of parameters used in the eclipse predictions, while bottom right gives the Moon's geocentric libration (optical + physical) at greatest eclipse.

1.3 Equidistant Conic Projection Maps of the Eclipse Paths

Figures 2.2 and 2.6 (Jan 15 annular) are maps that use an equidistant conic projection chosen to minimize distortion, and that isolate the land portions of the annular path. Curves of maximum eclipse are plotted and labeled at intervals of 30 min, while curves of constant eclipse magnitude appear at intervals of 0.1 magnitudes. A linear scale is included for estimating approximate distances (in kilometers). Within the northern and southern limits of the path of annularity, the outline of the antumbral shadow is plotted at intervals of 15 min. The Universal Time, the duration of annularity (in minutes and seconds), and the Sun's altitude are given at mid-eclipse for each shadow position.

1.4 Detailed Maps of the Eclipse Paths

The path of annularity or totality is plotted on a series of detailed maps appearing in Figures 2.3 to 2.5, 2.7 to 2.14 (Jan 15 annular), and Figures 3.2 to 3.6 (Jul 11 total). The maps were chosen to isolate small regions of each eclipse path over the entire land portion of the tracks and to include ocean sections containing islands. Curves of maximum eclipse are plotted at 5 min or shorter intervals along each path and labeled with the Universal Time, the central line duration of annularity or totality, and the Sun's altitude. The maps are constructed from the Digital Chart of the World (DCW), a digital database of the world developed by the U.S. Defense Mapping Agency (DMA). The primary sources of information for the geographic database are the Operational Navigation Charts (ONC) and the Jet Navigation Charts (JNC) developed by the DMA.

The scale varies from map to map depending partly on the population density and accessibility. The scale is adequate for showing the roads, villages, and cities required for eclipse expedition planning. The DCW database used for the maps was developed in the 1980s and contains place names in use during that period. Whenever possible, the DCW place names have been replaced with current names in use from the World Gazetteer at <http://www.world-gazetteer.com/>.

While Tables 2.1 to 2.6 (Jan 15 annular) and Tables 3.1 to 3.6 (Jul 11 total) deal with eclipse elements and specific characteristics of the path calculated at 5 min intervals, the northern and southern limits, as well as the central line of the path, are plotted using data generated at a higher cadence. These mapping data are available at the NASA Web sites for the 2010 annular eclipse <http://eclipse.gsfc.nasa.gov/SEmono/ASE2010/ASE2010.html>, and the 2010 total eclipse <http://eclipse.gsfc.nasa.gov/SEmono/TSE2010/TSE2010.html>.

Although no corrections have been made for center of figure or lunar limb profile, they have no observable effect at the scale of the maps. Atmospheric refraction has not been included, as it plays a significant role only at very low solar altitudes. The primary effect of refraction is to shift the path opposite to that of the Sun's local azimuth. This amounts to approximately 0.5° at the extreme ends, i.e., sunrise and sunset, of the umbral path. In any case, refraction corrections to the path are uncertain because they depend on the atmospheric temperature-pressure profile, which cannot be predicted in advance. A special feature of the maps are the curves of constant umbral (annular or total) eclipse duration which are plotted within the path at 1 or 2 min increments. These curves permit fast determination of approximate durations without consulting any tables.

Major highways are delineated in dark broad lines, but secondary and soft-surface roads are not distinguished, so caution should be used in this regard. If observations from the graze zones are planned, then the zones of grazing eclipse must be plotted on higher scale maps using graze path coordinates, which are available at the NASA Web sites for the 2010 annular eclipse <http://eclipse.gsfc.nasa.gov/SEmono/ASE2010/ASE2010.html>, and the 2010 total eclipse <http://eclipse.gsfc.nasa.gov/SEmono/TSE2010/TSE2010.html>. See Sect. 4.5 "Plotting Eclipse Paths on Maps" for sources and more information. The paths also show the curves of maximum eclipse at 5 min increments in Universal Time. The maps are also available at the NASA Web site for the 2010 solar eclipses.

1.5 Elements, Shadow Contacts, and Eclipse Path Tables

The geocentric ephemeris for the Sun and Moon, various parameters, constants, and the Besselian elements (polynomial form) are given in Tables 2.1 (Jan 15 annular) and 3.1 (Jul 11 total). The eclipse elements and predictions were derived from the DE200 and LE200 ephemerides (solar and lunar, respectively) developed jointly by NASA's Jet Propulsion Laboratory (JPL) and the U.S. Naval Observatory for use in the *Astronomical Almanac* beginning in 1984. Unless otherwise stated, all predictions are based on center of mass positions for the Moon and Sun with no corrections made for center of figure, lunar limb profile, or atmospheric refraction. The predictions depart from normal International Astronomical Union (IAU) convention through the use of a smaller constant for the mean lunar radius k for all umbral contacts (see Sect. 1.8 "Lunar Limb Profile"). Times are expressed in either Terrestrial Dynamical Time (TDT) or in Universal Time (UT), where the best value of ΔT (the difference between Terrestrial Dynamical Time and Universal Time) available at the time of preparation, is used.

The Besselian elements are used to predict all aspects and circumstances of a solar eclipse. The simplified geometry introduced by Bessel in 1824 transforms the orbital motions of the Sun and Moon into the position, motion, and size of the Moon's penumbral and umbral shadows with respect to a plane passing through Earth. This fundamental plane is constructed in an x–y rectangular coordinate system with its origin at Earth's center. The axes are oriented with north in the positive y direction and east in the positive x direction. The z-axis is perpendicular to the fundamental plane and parallel to the shadow axis.

The x and y coordinates of the shadow axis are expressed in units of the equatorial radius of Earth. The radii of the penumbral and umbral shadows on the fundamental plane are l_1 and l_2, respectively. The direction of the shadow axis on the celestial sphere is defined by its declination d and ephemeris hour angle μ. Finally, the angles that the penumbral and umbral shadow cones make with the shadow axis are expressed as f_1 and f_2, respectively. The details of actual eclipse calculations can be found in the *Explanatory Supplement* (Her Majesty's Nautical Almanac Office 1974) and *Elements of Solar Eclipses* (Meeus 1989).

From the polynomial form of the Besselian elements, any element can be evaluated for any time t_1 (in decimal hours) during the eclipse via the equation

$$a = a_0 + a_1 t + a_2 t^2 + a_3 t^3 \qquad (1)$$

(or $a = \sum [a_n t^n]$; $n = 0$ to 3),

where $a = x, y, d, l_1, l_2$, or μ; and $t = t_1 - t_0$ (decimal hours) and t_0 = 7.00 TDT (Jan 15 annular) or t_0 = 20.00 TDT (Jul 11 total).

The polynomial Besselian elements were derived from a least-squares fit to elements rigorously calculated at five separate times over a 6 h period centered at t_0.

Tables 2.2 (Jan 15 annular) and 3.2 (Jul 11 total) lists all external and internal contacts of penumbral and umbral shadows with Earth. They include TDT and geodetic coordinates with and without corrections for ΔT. The contacts are defined:

P1—Instant of first external tangency of penumbral shadow cone with Earth's limb (partial eclipse begins).
P2—Instant of first internal tangency of penumbral shadow cone with Earth's limb.
P3—Instant of last internal tangency of penumbral shadow cone with Earth's limb.
P4—Instant of last external tangency of penumbral shadow cone with Earth's limb (partial eclipse ends).
U1—Instant of first external tangency of umbral shadow cone with Earth's limb (annular or total eclipse begins).
U2—Instant of first internal tangency of umbral shadow cone with Earth's limb.
U3—Instant of last internal tangency of umbral shadow cone with Earth's limb.
U4—Instant of last external tangency of umbral shadow cone with Earth's limb (annular or total eclipse ends).

Similarly, the northern and southern extremes of the penumbral and umbral paths, and extreme limits of the central line are given. The IAU longitude convention is used throughout this publication (i.e., for longitude, east is positive and west is negative; for latitude, north is positive and south is negative).

The path of the umbral shadow is delineated at 5 min intervals (in Universal Time) in Tables 2.3 and 3.3. Coordinates of the northern limit, the southern limit, and the central line are listed to the nearest tenth of an arc minute (~185 m at the equator). The Sun's altitude, path width, and duration are calculated for the central line position. Tables 2.4 and 3.4 present a physical ephemeris for the umbral shadow at 5 min intervals in Universal Time. The central line coordinates are followed by the topocentric ratio of the apparent diameters of the Moon and Sun, the eclipse obscuration (defined as the fraction of the Sun's surface area occulted by the Moon), and the Sun's altitude and azimuth at that instant. The umbral path width, the umbral shadow's major and minor axes, and its instantaneous velocity with respect to Earth's surface are included. Finally, the central line duration of the annular or total phase is given.

Local circumstances for each central line position, listed in Tables 2.3 and 3.3, are presented in Tables 2.5 and 3.5, respectively. The first three columns give the Universal Time of maximum eclipse, the central line duration of annularity or totality, and the altitude of the Sun at that instant. The following columns list each of the four eclipse contact times followed by their related contact position angles and the corresponding altitude of the Sun. The four contacts identify significant stages in the progress of the eclipse. They are defined as follows:

First Contact: Instant of first external tangency between the Moon and Sun (partial eclipse begins).
Second Contact: Instant of first internal tangency between the Moon and Sun (annular or total eclipse begins).
Third Contact: Instant of last internal tangency between the Moon and Sun (annular or total eclipse ends).
Fourth Contact: Instant of last external tangency between the Moon and Sun (partial eclipse ends).

The position angles **P** and **V** (where **P** is defined as the contact angle measured counterclockwise from the equatorial *north* point of the Sun's disk, and **V** is defined as the contact angle measured counterclockwise from the local *zenith* point of the Sun's disk) identify the point along the Sun's disk where each contact occurs. Second and third contact altitudes are omitted because they are always within 1° of the altitude at maximum eclipse.

Tables 2.6 and 3.6 present topocentric values from the umbral path at maximum eclipse for the Moon's horizontal parallax, semi-diameter, relative angular velocity with respect to the Sun, and libration in longitude. The altitude and azimuth of the Sun are given along with the azimuth of the umbral path. The northern limit position angle identifies the point on the lunar disk defining the umbral path's northern limit. It is measured counterclockwise from the equatorial north point of the Moon. In addition, corrections to the path limits due to the lunar limb profile are listed (minutes of arc in latitude). The irregular profile of the Moon results in a zone of "grazing eclipse" at each limit, which is delineated by interior and exterior contacts of lunar features with the Sun's limb.

1.6 Local Circumstances Tables

Local circumstances for several hundred cities; metropolitan areas, and places are presented in Tables 2.7 to 2.15 (Jan 15 annular), and Table 3.7 (Jul 11 total). The tables give the local circumstances at each contact and at maximum eclipse for every location. (For partial eclipses, maximum eclipse is the instant when the greatest fraction of the Sun's diameter is occulted. For annular and total eclipses, maximum eclipse is the instant of mid-annularity or mid-totality.) The coordinates are listed along with the location's elevation (in meters) above sea level, if known. If the elevation is unknown (i.e., not in the database), then the local circumstances for that location are calculated at sea level. The elevation does not play a significant role in the predictions unless the location is near the umbral path limits or the Sun's altitude is relatively small (<10°).

The Universal Time of each contact is given to a tenth of a second, along with position angles **P** and **V** and the altitude of the Sun. The position angles identify the point along the Sun's disk where each contact occurs and are measured counterclockwise (i.e., eastward) from the north and zenith points, respectively. Locations outside the umbral path miss the annular or total eclipse and only witness first and fourth contacts. The Universal Time of maximum eclipse (either partial or total) is listed to a tenth of a second. Next, the position angles **P** and **V** of the Moon's disk with respect to the Sun are given, followed by the altitude and azimuth of the Sun at maximum eclipse. Finally, the corresponding eclipse magnitude and obscuration are listed. For umbral eclipses (both annular and total), the eclipse magnitude is identical to the topocentric ratio of the Moon's and Sun's apparent diameters.

Two additional columns are included if the location lies within the path of the Moon's umbral shadow. The "umbral depth" is a relative measure of a location's position with respect to the central line and path limits. It is a unit less parameter, which is defined as

$$u = 1 - (2\ x/W), \qquad (2)$$

where:

- u is the umbral depth,
- x is the perpendicular distance from the central line in kilometers, and
- W is the width of the path in kilometers.

The umbral depth for a location varies from 0.0 to 1.0. A position at the path limits corresponds to a value of 0.0, while a position on the central line has a value of 1.0. The parameter can be used to quickly determine the corresponding central line duration; thus, it is a useful tool for evaluating the trade-off in duration of a location's position relative to the central line. Using the location's duration and umbral depth, the central line duration is calculated as

$$D = d/[1 - (1-u)^2]^{1/2}, \qquad (3)$$

where:

- D is the duration of annularity or totality on the central line (in seconds),
- d is the duration of annularity or totality at location (in seconds), and
- u is the umbral depth.

The final column gives the duration of annularity or totality. The effects of refraction have not been included in these calculations, nor have there been any corrections for center of figure or the lunar limb profile.

Locations were chosen based on general geographic distribution, population, and proximity to the path. The primary source for geographic coordinates is *The New International Atlas* (Rand McNally 1991). Elevations for major cities were taken from *Climates of the World* (U.S. Dept. of Commerce, 1972). In this rapidly changing political world, it is often difficult to ascertain the correct name or spelling for a given location; therefore, the information presented here is for location purposes only and is not meant to be authoritative. Furthermore, it does not imply recognition of status of any location by the United States Government. Corrections to names, spellings, coordinates, and elevations should be forwarded to the authors in order to update the geographic database for future eclipse predictions.

1.7 Mean Lunar Radius

A fundamental parameter used in eclipse predictions is the Moon's radius k, expressed in units of Earth's equatorial radius. The Moon's actual radius varies as a function of position angle and libration because of the irregularity in the limb profile. From 1968 to 1980, the Nautical Almanac Office used two separate values for k in their predictions. The larger value ($k=0.2724880$), representing a mean over topographic features, was used for all penumbral (exterior) contacts and for annular eclipses. A smaller value ($k=0.272281$), representing a mean minimum radius, was reserved exclusively for umbral (interior) contact calculations of total eclipses (*Explanatory Supplement*, Her Majesty's Nautical Almanac Office, 1974). Unfortunately, the use of two different values of k for umbral eclipses introduces a discontinuity in the case of hybrid (annular-total) eclipses.

In 1982, the IAU General Assembly adopted a value of $k=0.2725076$ for the mean lunar radius. This value is now used by the Nautical Almanac Office for all solar eclipse predictions (Fiala and Lukac 1983) and is currently accepted as the best mean radius, averaging mountain peaks and low valleys along the Moon's rugged limb. The adoption of one single value for k eliminates the discontinuity in the case of hybrid eclipses and ends confusion arising from the use of two different values; however, the use of even the "best" mean value for the Moon's radius introduces a problem in predicting the true character and duration of umbral eclipses, particularly total eclipses.

During a total eclipse, the Sun's disk is completely occulted by the Moon. This cannot occur so long as any photospheric rays are visible through deep valleys along the Moon's limb (Meeus et al. 1966). The use of the IAU's mean k, however, guarantees that some annular or hybrid eclipses will be misidentified as total. A case in point is the eclipse of 1986 October 03. Using the IAU value for k, the *Astronomical Almanac* identified this event as a total eclipse of 3 s duration when it was, in fact, a beaded annular eclipse. Because a smaller value of k is more representative of the deeper lunar valleys and hence, the minimum solid disk radius, it helps ensure an eclipse's correct classification.

Of primary interest to most observers are the times when an umbral eclipse begins and when it ends (second and third contacts, respectively), and the duration of the umbral phase. When the IAU's value for k is used to calculate these times, they must be corrected to accommodate low valleys (total) or high mountains (annular) along the Moon's limb. The calculation of these corrections is not trivial, but is necessary, especially if one plans to observe near the path limits (Herald 1983). For observers near the central line of a total eclipse, the limb corrections can be more closely approximated by using a smaller value of k, which accounts for the valleys along the profile.

This publication uses the IAU's accepted value of $k=0.2725076$ for all penumbral (exterior) contacts. In order to avoid eclipse type misidentification and to predict umbral durations, which are closer to the actual durations at total eclipses, this document departs from IAU convention by adopting the smaller value of $k=0.272281$ for all umbral (interior) contacts. This is consistent with predictions in *Fifty Year Canon of Solar Eclipses: 1986–2035* (Espenak 1987) and *Five Millennium Canon of Solar Eclipses: –1999 to +3000* (Espenak and Meeus 2006). Consequently, the smaller k value produces shorter umbral durations and narrower paths for total eclipses when compared with calculations using the IAU value for k. Similarly, predictions using a smaller k value results in longer umbral durations and wider paths for annular eclipses than do predictions using the IAU's k value.

1.8 Lunar Limb Profile

Eclipse contact times, magnitude, and duration of annularity or totality all depend on the angular diameters and relative velocities of the Moon and Sun. Unfortunately, these calculations are limited in accuracy by the departure of the Moon's limb from a perfectly circular figure. The Moon's surface exhibits a dramatic topography, which manifests itself as an irregular limb when seen in profile. Most eclipse calculations assume some mean radius that averages high mountain peaks and low valleys along the Moon's rugged limb. Such an approximation is acceptable for many applications, but when higher accuracy is needed, the Moon's actual limb profile must be considered. Fortunately, an extensive body of knowledge exists on this subject in the form of Watts's limb charts (Watts 1963). These data are the product of a photographic survey of the marginal zone of the Moon and give limb profile heights with respect to an adopted smooth reference surface (or datum).

Analyses of lunar occultations of stars by Van Flandern (1970) and Morrison (1979) showed that the average cross section of Watts's datum is slightly elliptical rather than circular. Furthermore, the implicit center of the datum (i.e., the center of figure) is displaced from the Moon's center of mass.

In a follow-up analysis of 66,000 occultations, Morrison and Appleby (1981) found that the radius of the datum appears to vary with libration. These variations produce systematic errors in Watts's original limb profile heights that attain 0.4 arcsec at some position angles, thus, corrections to Watts's limb data are necessary to ensure that the reference datum is a sphere with its center at the center of mass.

The Watts charts were digitized by Her Majesty's Nautical Almanac Office (then in Herstmonceux, England), and transformed to grid-profile format at the U.S. Naval Observatory. In this computer readable form, the Watts limb charts lend themselves to the generation of limb profiles for any lunar libration. Ellipticity and libration corrections may be applied to refer the profile to the Moon's center of mass. Such a profile can then be used to correct eclipse predictions, which have been generated using a mean lunar limb.

Figures 2.15 and 3.7 give the limb profiles for the 2010 annular and total eclipses, respectively, for single location and time. The radial scale of the limb profiles (at bottom of each figure) is greatly exaggerated so that the true limb's departure from the mean lunar limb is readily apparent. The mean limb with respect to the center of figure of Watts's original data is shown (dashed curve) along with the mean limb with respect to the center of mass (solid curve). Note that all the predictions presented in this publication are calculated with respect to the latter limb unless otherwise noted. Position angles of various lunar features can be read using the protractor marks along the Moon's mean limb (center of mass). The position angles of second and third contact are clearly marked, as are the north pole of the Moon's axis of rotation and the observer's zenith at mid-totality. The dashed line with arrows at either end identifies the contact points on the limb corresponding to the northern and southern limits of the path. To the upper left of the profile, are the Sun's topocentric coordinates at maximum eclipse. They include the right ascension (*R.A.*), declination (*Dec.*), semi-diameter (*S.D.*), and horizontal parallax (*H.P.*) The corresponding topocentric coordinates for the Moon are to the upper right. Below and left of the profile are the geographic coordinates of the central line at selected locations, while the times of the four eclipse contacts at each location appear to the lower right. The limb-corrected times of second and third contacts are listed with the applied correction to the center of mass prediction.

Directly below the limb profile are the local circumstances at maximum eclipse. They include the Sun's altitude and azimuth, the path width, and umbral duration. The position angle of the path's northern-to-southern limit axis is *PA(N.Limit)* and the angular velocity of the Moon with respect to the Sun is *A.Vel.(M:S)*. At the bottom left are a number of parameters used in the predictions, and the topocentric lunar librations appear at the lower right.

In investigations where accurate contact times are needed, the lunar limb profile can be used to correct the nominal or mean limb predictions. For any given position angle, there will be a high mountain (annular eclipses) or a low valley (total eclipses) in the vicinity that ultimately determines the true instant of contact. The difference, in time, between the Sun's position when tangent to the contact point on the mean limb and tangent to the highest mountain (annular) or lowest valley (total) at actual contact is the desired correction to the predicted contact time. On the exaggerated radial scale of Figures 2.15

and 3.7, the Sun's limb can be represented as an epicyclic curve that is tangent to the mean lunar limb at the point of contact and departs from the limb by h through

$$h = S(m-1)(1-\cos[C]), \qquad (9)$$

where:
 h is the departure of Sun's limb from mean lunar limb,
 S is the Sun's semi-diameter,
 m is the eclipse magnitude, and
 C is the angle from the point of contact.

Herald (1983) takes advantage of this geometry in developing a graphic procedure for estimating correction times over a range of position angles. Briefly, a displacement curve of the Sun's limb is constructed on a transparent overlay by way of equation (9). For a given position angle, the solar limb overlay is moved radially from the mean lunar limb contact point until it is tangent to the lowest lunar profile feature in the vicinity. The solar limb's distance **d** (in arc seconds) from the mean lunar limb is then converted to a time correction δ by

$$\delta = d/v \cos[X - C], \qquad (10)$$

where:
 δ is the correction to contact time (in seconds),
 d is the distance of solar limb from Moon's mean limb (in arc seconds),
 v is the angular velocity of the Moon with respect to the Sun (arc seconds per second),
 X is the central line position angle of the contact, and
 C is the angle from the point of contact.

This operation may be used for predicting the formation and location of Baily's beads. When calculations are performed over a large range of position angles, a contact time correction curve can then be constructed.

Because the limb profile data are available in digital form, an analytical solution to the problem is possible that is quite straightforward and robust. Curves of corrections to the times of second and third contact for most position angles have been computer generated and are plotted in Figures 2.15 and 3.7. The circular protractor scale at the center represents the nominal contact time using a mean lunar limb. The departure of the contact correction curves from this scale graphically illustrates the time correction to the mean predictions for any position angle as a result of the Moon's true limb profile. Time corrections external to the circular scale are added to the mean contact time; time corrections internal to the protractor are subtracted from the mean contact time. The magnitude of the time correction at a given position angle is measured using any of the four radial scales plotted at each cardinal point.

2. Annular Solar Eclipse of 2010 Jan 15

2.1 Introduction

On Friday, 2010 January 15, an annular eclipse of the Sun is visible from within a 300 km wide track that traverses half of Earth. The path of the Moon's antumbral shadow begins in Africa and passes through Chad, Central African Republic, Democratic Republic of the Congo, Uganda, Kenya, and Somalia. After leaving Africa, the path crosses the Indian Ocean where the maximum duration of annularity reaches 11 min 08 s (Espenak 1987). The central path then continues into Asia through Bangladesh, India, Burma (Myanmar), and China. A partial eclipse is seen within the much broader path of the Moon's penumbral shadow, which includes eastern Europe, most of Africa, Asia, and Indonesia (Figure 2.1).

2.2 Antumbral Path and Visibility

Earth reaches perihelion on Jan 03, just 12 days before the annular eclipse, so the Sun is nearly at its maximum apparent diameter. The Moon passes through apogee on Jan 17 (01:41 UT) so it is close to its minimum apparent diameter during the eclipse. The combination of these factors results in an unusually wide path of annularity.

The track of the Moon's shadow begins in western-most Central African Republic (Figure 2.2) at 05:14 UT. The eclipse path is 371 km wide at its start as the antumbra quickly travels east-southeast.

The northern edge of the path begins in southern Chad (Figure 2.3), but quickly crosses into the Central African Republic (CAR). Bangui, the capital of, and the largest city in, the CAR, lies 150 km south of the central line. Its 500,000 inhabitants witness a 4 min 0 s annular eclipse with the Sun just 4° above the eastern horizon at 05:17 UT. To the north, the duration on the central line is 7 min 18 s. At this instant, the enormous antumbral shadow is racing across the African continent with a velocity exceeding 10 km/s, but its speed is dropping fast.

When the central line crosses into the Democratic Republic of the Congo (DRC) at 05:19 UT, the duration hits 7 min 30 s and the Sun's altitude is 8°. DRC's capital city, Kisangani, is 115 km south of the southern limit (Figure 2.4). Its population of more than 7 million must be content with a partial eclipse of 0.88.

Continuing eastward, the shadow enters Uganda and engulfs Lake Albert and Kabarega National Park (05:23 UT). The central line duration now tops 8 min, the Sun stands 17° above the horizon, and the antumbral velocity is 2.3 km/s. Uganda's capital, Kampala, with a population of 1.2 million, is 60 km south of the central line but still deep in the annular path. The central phase of the eclipse lasts 7 min 36 s in Kampala, while the central line duration is 8 min 12 s. The northern half of Lake Victoria, Africa's largest lake, also extends into the antumbral track.

At 05:27 UT, the central line exits Uganda and enters Kenya (Figure 2.5). Many of Kenya's national parks and wildlife reserves fall within the annular path including Mt. Elgon, Keno Valley, Samburu, Mt. Kenya, Lake Nakuru, Meru, Kora, and Tsavo East. The southern edge of the path bisects Masai Mara Game Reserve. Nairobi, Kenya's capital of 2.9 million people, stands within 100 km of the central line and experiences 6 min 54 s of annularity. Due north of Nairobi on the central line, the duration is 8 min 32 s, the Sun is 25° above the horizon, and the antumbral velocity is 1.5 km/s (05:30 UT). The long duration implies that the Moon appears considerably smaller than the Sun. Indeed, the lunar diameter is only 0.912 of the Sun's disk, obscuring just 0.832 of the Sun's surface area. Most people will not even be aware of any local darkening at maximum eclipse unless it is pointed out.

The Moon's shadow continues through Kenya as its northern half briefly enters southernmost Somalia. As the central line exits Kenya and heads across the Indian Ocean (05:37 UT), its annular duration is 1 s shy of 9 min. For the next 2 h, the antumbra crosses the Indian Ocean, its course slowly curving from east-southeast to northeast. The Seychelles lie just outside of the track and get a deep partial eclipse of magnitude 0.907 at 06:09 UT.

The instant of greatest eclipse occurs at 07:06:33 UT (latitude 01° 37′N, longitude 69° 17′E) when the axis of the Moon's shadow passes closest to the center of Earth (gamma[1] = +0.40016). The maximum duration of annularity here is 11 min 08 s, the Sun's altitude is 66°, the path width is 333 km, and the antumbra's velocity is 0.46 km/s. Although the Moon's relative diameter to the Sun (0.919) is slightly larger than it was in Kenya, its relative motion is considerably smaller due in larger part to Earth's rotation along with the antumbral, making the central duration several minutes longer. This is, in fact, the longest annular eclipse of the 3rd Millennium—its duration will not be exceeded until the year 3043.

Although the eclipse is half over, it must now traverse southern Asia before reaching its terminus (Figure 2.6). The next landfall occurs at about 07:26 UT when the shadow sweeps over the Maldive Islands (Figure 2.7). The capital city of Malé is just 22 km from the central line so it experiences a long annular duration of 10 min 45 s. This is the longest duration of any city having an international airport in the eclipse track.

The path finally reaches Asia as the central line passes directly between the southern tip of India and northern Sri Lanka. Both regions lie within the path (Figure 2.8) although Sri Lanka's capital city Colombo (pop. 640,000) is outside the southern limit and receives a partial eclipse of magnitude 0.902 at 07:49 UT.

Leaving the Indian subcontinent, the antumbra crosses the Bay of Bengal before its central axis reaches the southwest coast of Burma (Myanmar) at 08:33 UT (Figure 2.9). The central line duration is 08 m 47 s, the solar altitude is 34°, the path width is 834 km, and the velocity is 1.1 km/s. The northern edge of the shadow briefly enters Bangladesh and southeastern

---

1 Gamma is the perpendicular distance of the Moon's shadow axis from Earth's center in units of equatorial Earth radii. It is measured when the distance to the geocenter reaches a minimum (i.e., instant of greatest eclipse).

India before returning completely into Burma. Mandalay is 75 km south of the central line where its 0.9 million inhabitants experience a 7 min 35 s annular phase centered at 08:38 UT.

At 08:41 UT, the antumbra's central line enters China, the largest and final country in the track. The shadow crosses the southern foot of the Himalayas through Yunnan and Sichuan provinces (Figures 2.10 and 2.11). Chongqing (urban population 2.2 million) is the largest and most populous of China's four provincial-level municipalities. It lies directly on the central line where the duration is 7 min 50 s and the Sun's altitude is 15°.

As the curvature of Earth brings the planet's surface farther from the Moon and at an increasingly oblique angle, the duration of annularity and the Sun's altitude decrease, while the antumbra's ground velocity increases. Racing through parts of Shaanxi, and Hubei provinces (Figures 2.12 and 2.13), the northern antumbra engulfs Zhenzhou, the capital city of Henan province. The Sun is 7° during a 4 min 30 s annular phase. On the central line 145 km to the southeast, the duration is still 7 min 27 s while the shadow's velocity is 6.6 km/s.

In its final moments, the antumbra enters Shandong province and the lower reaches of the Yellow River (Figure 2.14). The track travels down the Shandong Peninsula where it ends as the Moon's shadow lifts off Earth (08:59 UT). During the course of its 3 h 45 min trajectory, the antumbra's track is approximately 12,900 km long, which covers 0.87% of Earth's surface area. It will be 29 months before the next annular solar eclipse occurs on 2012 May 20.

2.3 Maps of the Annular Eclipse Path

Maps of the Jan 15 annular eclipse path are given in Figures 2.1 through 2.14. Figure 2.1 is an orthographic projection map of Earth showing the path of penumbral (partial) and antumbral (annular) eclipse. The limits of the Moon's penumbral shadow define the region of visibility of the partial eclipse. The much narrower path of the antumbral shadow defines the zone where the annular eclipse is visible. For a more detailed description of Figure 3.1, see Section 1.2.

Figures 2.2 through 2.14 offer more detailed maps of the entire land portion of the path of annularity, as well as ocean sections of the track containing islands. A complete description of these figures can be found in Sections 1.3 and 1.4.

2.4 Annular Eclipse Elements and Path Tables

Tables 2.1 through 2.7 give elements for the eclipse, as well as basic characteristics of the annular path. The geocentric ephemeris for the Sun and Moon, various parameters, constants, and the Besselian elements (polynomial form) are found in Table 2.1. All external and internal contacts of penumbral and antumbral shadows with Earth are listed in Table 2.2. They include TDT and geodetic coordinates with and without corrections for ΔT.

The path of the antumbral shadow is delineated at 5 min intervals (in Universal Time) in Table 2.3. Coordinates of the northern limit, the southern limit, and the central line are listed along with the Sun's altitude, path width, and central line duration of annularity. Table 2.4 presents a physical ephemeris for the antumbral shadow and includes the topocentric ratio of the Moon's and Sun's apparent diameters, the eclipse obscuration, the path width, the dimensions of the antumbral shadow, and its ground velocity.

Table 2.5 gives the local circumstances for each central line position listed in Tables 2.3 and 2.4. Table 2.6 presents topocentric values from the central path for the Moon's horizontal parallax, semi-diameter, relative angular velocity with respect to the Sun, and libration in longitude. In addition, corrections to the path limits due to the lunar limb profile are listed. A detailed description of these tables can be found in Section 1.5.

2.5 Annular Eclipse Local Circumstances Tables

Local circumstances for many cities, metropolitan areas, and places in Africa, Europe, the Middle East, and Asia are presented in Tables 2.7 to 2.15. The tables give the local circumstances at each contact and at maximum eclipse for every location. The coordinates are listed along with the location's elevation (in meters) above sea level. The Universal Time of each contact is given to a tenth of a second, along with position angles P and V and the altitude of the Sun. Two additional columns are included if the location lies within the path of the Moon's antumbral shadow. The "umbral depth" is a relative measure of a location's position with respect to the central line and path limits. The last column gives the duration of annularity. For more information about these tables, see Section 1.6.

2.6 Annular Eclipse Lunar Limb Profile

Along the 2010 annular eclipse path, the Moon's topocentric libration (physical plus optical) in longitude ranges from $l = +2.4°$ to $l = +0.6°$; thus, a limb profile with the appropriate libration is required in any detailed analysis of contact times, central durations, etc. A profile with an intermediate value, however, is useful for planning purposes and may even be adequate for most applications. The lunar limb profile presented in Figure 2.15 includes corrections for center of mass and ellipticity (Morrison and Appleby 1981). It is generated for 7:00 UT, which corresponds to the integral hour nearest greatest eclipse. The antumbral shadow is then located in the Indian Ocean at latitude 00° 49.9′N and longitude 67° 52.1′E. The Moon's topocentric libration is $l = +1.58°$, and the topocentric semi-diameters of the Sun and Moon are 975.6 and 896.6 arcsec, respectively. The Moon's angular velocity with respect to the Sun is 0.236 arcsec/s.

The times of the four eclipse contacts from this location appear to the lower right in Figure 2.15. The limb-corrected times of second and third contacts are listed with the applied correction to the center of mass prediction. The time correction curves can be used for estimating corrections to the times of second and third contacts as a function of the position angle of the contact. More information on this topic and a detailed description of the limb profile figure can be found in Section 1.8.

2.7 Weather Prospects for the Annular Eclipse

2.7.1 Introduction

During the annular eclipse of 2010, Earth intercepts the Moon's shadow during the dry monsoons of January. In many places along the path, the weather prospects are almost guaranteed to be excellent. Alas, the political and social climates are major concerns, and these factors will have to be considered when selecting an observing site.

In January, the Sun is far to the south and winter is in its depths over the Northern Hemisphere. Over tropical regions of Africa and Southeast Asia, winds turn to the northwest, tapping the warm dry air of the Sahara in Africa, or the dry cool air descending from the Himalayas in India and Burma. Sunshine dominates in most regions, interrupted only where high terrain or nearby moisture sources provide the forcing necessary to produce clouds or where the path makes a feint toward the tropical rain forests of Africa. While moisture is abundant over the Indian Ocean, the strong inversions created by air subsiding aloft suppresses the development of clouds and precipitation. On the north side of the Himalayas, the eclipse track encounters China's winter season, with air masses too cold to support extensive cloudiness.

2.7.2 Africa

Between the arid Sahara Desert and the equatorial rain forest, lies a climatic region known as the wet and dry tropics and it is through this zone that the Moon's antumbral shadow makes its sunrise contact and early passage across the globe. Earth's weather equator or Intertropical Convergence Zone (ITCZ) has migrated southward to its winter latitudes, permitting the drier continental air from the Sahara to spread toward the eclipse path. It is not a complete victory for the desert air, as the track skips along the boundary between the dry and wet zones and even dips into a nose of moist air over the Democratic Republic of the Congo (DRC) and Uganda before crossing back into the sunnier climate of Kenya (Figure 2.2).

At Bangui in the Central African Republic, average January precipitation amounts to only 25 mm and measurable rains fall on only three days of the month. Sunshine is an encouraging 59% of the maximum possible at Bangui, and probably as much as 10% higher at the beginning of the track over the northwestern part of the country. Farther along, where the path dips into the DRC, the closer proximity of the ITCZ increases average cloudiness to 65% to 75% (Kisangani: Table 2.16) and a sunshine frequency of only 50%.

As the lunar shadow's path reaches the DRC–Uganda border (Figure 2.4), the terrain changes dramatically, rising from the flat sierra of the DRC to the elevated plateau of Uganda and western Kenya. It is a region of complex terrain, with a highly variable climatology that depends on elevation, airflow, and moisture supply, all of which change in small scale. In the eastern DRC, the Mitumba Mountains and their northward extension are a volcanic chain that rises above 5000 m at Mount Stanley and at Margherita Peak and brings a heavier cloudiness than the lower plains. Immediately east of the mountains, along the DRC–Uganda border, the Western Rift Valley carves a deep trench in the terrain to hold Lake Edward and Lake Albert. Once past the Rift Valley, the terrain becomes more sedate, and the eclipse path transits the flat uplands of Uganda where moisture is readily available because of the swampy terrain and the presence of Lake Victoria, the largest lake in Africa. Cloudiness is highly variable (Table 2.16), with average amounts ranging from 54% to 77% depending on the proximity of Lake Victoria. Masindi, in central Uganda and well removed from Lake Victoria, seems to offer the best weather with an average cloudiness of 54% and perhaps a 70% probability of sunny weather on eclipse day. Entebbe Airport, along the shores of Lake Victoria, records a much higher cloudiness, but its 62% frequency of sunshine is also promising.

Leaving Uganda, the eclipse path moves back into the rougher terrain that dominates western Kenya (Figure 2.5). This is a region of tall volcanoes, framed by Mount Elgon (4321 m) in the west and Mount Kenya (5199 m) in the east, and cut down the middle by the Eastern Rift Valley. The position of the ITCZ lies well to the south of Kenya in January and the country offers the best prospects for sunshine in Africa, in spite of its rugged topography. Average cloudiness is below 50% at Eldoret and Kitale, and under 60% in the rest of the western highlands. The percent of maximum possible sunshine rises to a surprising 76% at Nairobi and is probably even higher at Eldoret. Without doubt, there will be some eclipse-watchers ensconced on Mount Kenya, though there are many other sites with high sunshine probabilities. The secret is to select a site in a valley behind one of the many mountains, so that the prevailing easterly winds (that is, winds blowing toward the west) must flow downhill to an eclipse-watcher's encampment. A quick look at a topographic map will identify suitable locations including Nairobi, Nakuru, Kisumu, Eldoret, and Kitale.

East of Nairobi, the eclipse moves onto the coastal plains of Kenya and southern Somalia. The north limit is considerably drier than the south, but the plains are only a little cloudier than the Kenyan highlands. The explanation lies in the pattern of the wind flow and the cool ocean currents, which together conspire to reduce the production of convective clouds in the region. Temperatures and humidity are uncomfortable through the lowlands, with daytime highs climbing into the mid-30s (°C), over 85° F, and overnight lows only falling to the mid-20s (°C).

Falling temperatures during the eclipse may cause the formation of fog and low clouds in the high humidity of lowland regions, though these types of clouds are usually rare in other circumstances. Eclipse sites should avoid rising terrain where the winds blow upslope. This is easier said than done, as the winds across Africa are typically light in the season, and the winds themselves may change with the eclipse cooling.

2.7.3 India and Sri Lanka

The eclipse path continues to head slowly southward as it leaves the African coast, leaving its northward turn until about halfway to India. This trend keeps the track on the northern margins of the ITCZ where the highest frequency of convective cloudiness occurs. At its most southerly extent, near the Seychelles Islands, the path encounters the most unfavorable climatology, with a mean cloud cover of more than 70% and a sunshine frequency of only 40%. By the time the track reaches the Maldives (Figure 2.7) and its capital Malé (23 km south of the central line), cloudiness has dropped to only 55% and the daily sunshine averages 2/3 of the amount possible.

A little farther along, past Minicoy, the antumbral path reaches the southern tip of India and the states of Kerala and Tamil Nadu (Figure 2.8). Landfall is near Thiruvananthapuram (Trivandrum), where the amount of sunshine climbs to 72%. The dry winter season over southern India is not as cloud-free as farther north, but the general conditions here are slightly more favorable than in Kenya. The path through Kerala and Tamil Nadu gives a high probability of sunshine on eclipse day. India might be the best location from which to observe at the north limit, where beading phenomena are at their best. Climate statistics for Madurai and Cuddalore in Tamil Nadu (Table 2.16) characterize the north limit of the shadow track; there is no climate information near the north limit in Kerala, but Thiruvananthapuram is probably an accurate reflection of the area. From the cloud statistics, the east coast seems more suitable than the west. As in Africa, the formation of fog is a possibility as the ground cools during the approach of the shadow, but this seems unlikely if winds are light or not blowing in an upslope direction.

Sri Lanka also offers favorable weather prospects, but the political situation makes it all but impossible to observe from the island.

2.7.4 Burma and Bangladesh

North-central Burma is a low landscape lying between two mountain chains, the Arakan Yoma inland from the coast of the Bay of Bengal, and the Shan Plateau in the west, which blends gradually into the Himalayan Mountains to the north. During January, the northeast monsoon is entrenched across the region, and the flow of air into the lowlands, though much modified by its passage southward across the subtropical landscape, retains the dry and stable character of its source region in Mongolia. Because of the Himalayas, cold Siberian air cannot reach Burma directly, but must first run through southern China and across Vietnam and Laos where the terrain is lower and where heat and moisture are added to the lower layers of the atmosphere. The region of the eclipse track, being rather well protected by the eastern mountains, enjoys a nearly rain-free climatology and abundant sunshine. Burma has the best weather prospects of any eclipse site along the path of the Moon's antumbral shadow.

Mandalay, in the Burma lowlands along the Irrawaddy River and south of the central line, offers more than 80% of the sunshine amount possible on a January day. The average cloud amount is a minuscule 15%—equal to the Sahara on its best days. January rainfall, though not completely absent, is only a few millimeters. Satellite observations of cloud cover (Figure 2.16) place the area around Mandalay among the sunniest parts of the globe in January.

The dry northeast monsoon brings favorable weather to all of Burma, even in the humid air along the Bay of Bengal coast where Sittwe has nearly the same cloud and sunshine characteristics as Mandalay. Mountains still have an influence, however, as Mindat, in the Arakan Yoma, and Lashio in the midst of the Shan Plateau, have a higher, though still-excellent, average cloudiness of around 30%. Bad weather, when it comes, is usually formed on the weak front that develops between the modified polar air from the north and intrusion of maritime tropical air from the equator. The frontal zone typically lies north of the eclipse track, but makes occasional excursions southward whenever a fresh surge of Mongolian air reaches the region. January is well outside the cyclone season.

In Burma, a location north (for the central line) or south (for the beaded zone) of Mandalay offers some of the best prospects. Highway 1, from Rangoon to Mandalay and beyond, provides convenient access to the width of the track.

The northern side of the eclipse path can also be reached at Cox's Bazarre in Bangladesh (Figure 2.9). One of the most-visited tourist locations in Bangladesh, it lies 27 km inside the north limit and boasts the same sunny character as Sittwe to the south. Its average cloud amount is even lower than Mandalay.

2.7.5 China

The accelerating shadow makes its last dash across Earth in China, moving steadily northeastward to its sunset ending in the Yellow Sea southeast of Beijing (Figure 2.6). The first part of the trek is across the rough terrain of Yunnan and Sichuan provinces, crossing the southern foot of the Himalayas until finally reaching the flat plains of central China (Figure 2.10). In January, the Siberian high is at its strongest and coldest, bringing steady northeast winds to the country. This is the season of the winter monsoon, with frequent southward-moving surges of cold air. These frontal outbreaks occur at a frequency of about one per week and typically involve the development of a low-pressure system along the front that brings extensive cloudiness and precipitation—a pattern that will be familiar to North American and European visitors. The cold air is relatively shallow and does not penetrate readily into the high terrain in Yunnan, which retains the sunny character of Burma.

The climate statistics in Table 2.16 show a steady decline in the percent of possible sunshine along the path, from 74% in Tengchong to 50% at Zengzhou. Central China, between Chongqing and Zengzhou, has the cloudiest weather according to the satellite observations (Figure 2.16). Farther northeast, where the dry, cold Siberian air has more influence, the weather prospects actually improve modestly as the eclipse track reaches its sunset terminus (Figure 2.14). This is a region that has the unique distinction of being visited by three eclipses in three years—two totals and now an annular. A fortunate observer at Xiangfan would have to travel no more than 100 km to see all three.

Temperatures decline along with the weather prospects across China, but it is only northeast of Chongqing that freezing weather becomes a serious possibility, and then mostly at night.

Along with the colder weather comes the possibility of snow, but most travelers from the Northern Hemisphere will find that the weather in China is less severe than that at home.

2.7.6 Summary

Thanks to the dry winter season, good weather can be found across much of Africa, the Indian Ocean, India, Burma, and southwest China. Mountains should be avoided, but the moist air of the Indian Ocean is not inclined to make clouds and so, only the humidity will interfere with observations of the eclipse. Southern India is attractive and central Burma offers the best weather of all. Winds are generally light, fog is rare (though it may form during the eclipse), and temperatures on the plateaus are tolerable while those near sea level are hot.

No region is without its cloudy days, however, and quick movement to a sunnier location is not easily done in rugged terrain or where roads are limited, so the best advice is to pick carefully at the start and hope for "climatological" weather on eclipse day.

TABLE 2.1

ELEMENTS OF THE ANNULAR SOLAR ECLIPSE OF 2010 JANUARY 15

```
Equatorial Conjunction:     07:21:27.24 TDT      J.D. = 2455211.806565
 (Sun & Moon in R.A.)      (=07:20:21.20 UT)

Ecliptic Conjunction:       07:12:28.46 TDT      J.D. = 2455211.800329
 (Sun & Moon in Ec. Lo.)   (=07:11:22.41 UT)

    Instant of              07:07:39.03 TDT      J.D. = 2455211.796980
Greatest Eclipse:          (=07:06:32.99 UT)
```

Geocentric Coordinates of Sun & Moon at Greatest Eclipse (JPL DE200/LE200):

```
Sun:     R.A. = 19h47m51.053s       Moon:    R.A. = 19h47m25.329s
         Dec. =-21°07'38.69"                 Dec. =-20°46'54.79"
  Semi-Diameter =   16'15.54"        Semi-Diameter =    14'44.35"
   Eq.Hor.Par. =      08.94"          Eq.Hor.Par. =  0°54'05.35"
       Δ R.A. =     10.744s/h             Δ R.A. =    122.609s/h
       Δ Dec. =       27.56"/h            Δ Dec. =     480.39"/h
```

```
Lunar Radius    k1 = 0.2725076 (Penumbra)     Shift in      Δb =  0.00"
 Constants:     k2 = 0.2722810 (Umbra)      Lunar Position: Δl =  0.00"

Geocentric Libration:   l =   1.5°    Brown Lun. No. = 1077
 (Optical + Physical)   b =  -0.5°    Saros Series  = 141 (23/70)
                        c =  -8.8°          nDot = -26.00 "/cy**2
```

Eclipse Magnitude = 0.91903   Gamma = 0.40016   ΔT =   66.0 s

Polynomial Besselian Elements for: 2010 Jan 15   07:00:00 TDT   (=t_0)

```
  n       x           y           d          l1          l2          μ

  0  -0.1732440   0.3664046 -21.1292992   0.5746956   0.0283960  282.671112
  1   0.4845213   0.1404923   0.0073072   0.0000372   0.0000370   14.997591
  2  -0.0000371   0.0001170   0.0000056  -0.0000099  -0.0000099    0.000002
  3  -0.0000054  -0.0000017   0.0000000   0.0000000   0.0000000    0.000000

            Tan f1 = 0.0047545    Tan f2 = 0.0047308
```

At time t_1 (decimal hours), each Besselian element is evaluated by:

$$a = a_0 + a_1 * t + a_2 * t^2 + a_3 * t^3 \quad \text{(or } a = \sum [a_n * t^n]; n = 0 \text{ to } 3)$$

where: a = x, y, d, l_1, l_2, or μ
          $t = t_1 - t_0$ (decimal hours) and t_0 = 7.00 TDT

The Besselian elements were derived from a least-squares fit to elements calculated at five uniformly spaced times over a 6-hour period centered at t_0. Thus, they are valid over the period $4.00 \le t_1 \le 10.00$ TDT.

All times are expressed in Terrestrial Dynamical Time (TDT).

Saros Series 141: Member 23 of 70 eclipses in series.

TABLE 2.2

SHADOW CONTACTS AND CIRCUMSTANCES
ANNULAR SOLAR ECLIPSE OF 2010 JANUARY 15

$$\Delta T = 66.0 \text{ s}$$
$$= 000°16'33.3"$$

		Terrestrial Dynamical Time h m s	Latitude	Ephemeris Longitude†	True Longitude*
External/Internal Contacts of Penumbra:	P_1	04:06:33.5	01°19.3'S	030°10.3'E	030°26.8'E
	P_2	06:51:12.8	49°46.0'N	016°42.4'E	016°58.9'E
	P_3	07:23:43.6	68°44.1'N	078°14.4'E	078°31.0'E
	P_4	10:08:41.1	28°48.1'N	107°54.8'E	108°11.4'E
Extreme North/South Limits of Penumbral Path:	N_1	06:38:30.2	55°02.1'N	026°15.4'E	026°31.9'E
	S_1	05:22:22.9	23°58.0'S	001°49.9'E	002°06.4'E
	N_2	07:36:14.0	68°32.5'N	057°41.0'E	057°57.6'E
	S_2	08:53:00.5	06°14.6'N	136°39.6'E	136°56.2'E
External/Internal Contacts of Umbra:	U_1	05:15:00.9	06°20.1'N	016°01.9'E	016°18.4'E
	U_2	05:22:21.8	07°39.7'N	014°43.0'E	014°59.5'E
	U_3	08:52:46.3	37°29.1'N	121°54.8'E	122°11.4'E
	U_4	09:00:09.7	36°12.5'N	120°52.3'E	121°08.8'E
Extreme North/South Limits of Umbral Path:	N_1	05:19:34.2	08°37.1'N	015°47.6'E	016°04.1'E
	S_1	05:17:54.1	05°21.4'N	014°55.6'E	015°12.2'E
	N_2	08:55:34.1	38°24.0'N	120°37.1'E	120°53.7'E
	S_2	08:57:16.1	35°16.0'N	122°10.1'E	122°26.7'E
Extreme Limits of Central Line:	C_1	05:18:40.7	06°58.7'N	015°22.1'E	015°38.6'E
	C_2	08:56:28.6	36°49.6'N	121°24.3'E	121°40.9'E
Instant of Greatest Eclipse:	G_0	07:07:39.0	01°37.4'N	069°00.9'E	069°17.4'E

Circumstances at
Greatest Eclipse: Sun's Altitude = 66.4°   Path Width = 333.1 km
                     Sun's Azimuth = 164.9°   Central Duration = 11m07.8s

---

† Ephemeris Longitude is the terrestrial dynamical longitude assuming a uniformly rotating Earth.
* True Longitude is calculated by correcting the Ephemeris Longitude for the non-uniform rotation of Earth.
  (T.L. = E.L. + 1.002738*ΔT/240, where ΔT(in seconds) = TDT - UT)

Note: Longitude is measured positive to the East.

Because ΔT is not known in advance, the value used in the predictions is an extrapolation based on pre-2009 measurements. The actual value is expected to fall within ±0.3 seconds of the estimated ΔT used here.

TABLE 2.3

PATH OF THE ANTUMBRAL SHADOW
ANNULAR SOLAR ECLIPSE OF 2010 JANUARY 15

$\Delta T = 66.0$ s

Universal Time	Northern Limit Latitude	Northern Limit Longitude	Southern Limit Latitude	Southern Limit Longitude	Central Line Latitude	Central Line Longitude	Sun Alt °	Path Width km	Central Durat.
Limits	08°37.1'N	016°04.1'E	05°21.4'N	015°12.2'E	06°58.7'N	015°38.6'E	0	371	07m09.4s
05:20	05°36.1'N	023°55.2'E	01°10.3'N	026°38.3'E	03°16.3'N	025°32.1'E	11	360	07m41.3s
05:25	02°54.0'N	031°38.9'E	00°51.6'S	032°56.7'E	00°58.9'N	032°22.3'E	19	354	08m09.8s
05:30	01°29.3'N	036°16.1'E	02°03.8'S	037°14.5'E	00°18.8'S	036°48.0'E	25	351	08m31.7s
05:35	00°33.2'N	039°47.9'E	02°52.9'S	040°39.0'E	01°11.1'S	040°15.3'E	30	349	08m50.8s
05:40	00°06.2'S	042°43.2'E	03°27.5'S	043°31.7'E	01°48.0'S	043°08.9'E	34	347	09m08.1s
05:45	00°34.2'S	045°14.8'E	03°51.7'S	046°02.8'E	02°14.0'S	045°39.9'E	37	347	09m24.0s
05:50	00°53.5'S	047°29.2'E	04°07.8'S	048°18.0'E	02°31.6'S	047°54.5'E	41	346	09m38.7s
05:55	01°05.8'S	049°30.6'E	04°17.4'S	050°21.0'E	02°42.5'S	049°56.6'E	44	346	09m52.3s
06:00	01°12.4'S	051°21.8'E	04°21.4'S	052°14.2'E	02°47.8'S	051°48.6'E	47	346	10m04.9s
06:05	01°14.0'S	053°04.7'E	04°20.6'S	053°59.4'E	02°48.2'S	053°32.5'E	49	346	10m16.4s
06:10	01°11.3'S	054°40.7'E	04°15.5'S	055°37.9'E	02°44.2'S	055°09.7'E	52	345	10m26.9s
06:15	01°04.6'S	056°11.0'E	04°06.6'S	057°10.8'E	02°36.4'S	056°41.2'E	54	345	10m36.3s
06:20	00°54.4'S	057°36.4'E	03°54.1'S	058°38.8'E	02°25.1'S	058°07.9'E	56	345	10m44.5s
06:25	00°40.9'S	058°57.6'E	03°38.4'S	060°02.7'E	02°10.5'S	059°30.4'E	58	344	10m51.7s
06:30	00°24.4'S	060°15.3'E	03°19.6'S	061°23.0'E	01°52.8'S	060°49.3'E	60	344	10m57.7s
06:35	00°04.9'S	061°30.0'E	02°58.0'S	062°40.2'E	01°32.3'S	062°05.2'E	61	343	11m02.5s
06:40	00°17.3'N	062°42.0'E	02°33.6'S	063°54.7'E	01°09.0'S	063°18.4'E	63	341	11m06.2s
06:45	00°42.2'N	063°51.8'E	02°06.6'S	065°06.9'E	00°43.1'S	064°29.4'E	64	340	11m08.8s
06:50	01°09.6'N	064°59.8'E	01°37.0'S	066°17.2'E	00°14.6'S	065°38.4'E	65	339	11m10.3s
06:55	01°39.6'N	066°06.2'E	01°04.9'S	067°25.8'E	00°16.4'N	066°45.9'E	66	337	11m10.8s
07:00	02°12.0'N	067°11.4'E	00°30.5'S	068°33.0'E	00°49.9'N	067°52.1'E	66	335	11m10.1s
07:05	02°47.0'N	068°15.7'E	00°06.4'N	069°39.2'E	01°25.8'N	068°57.3'E	66	334	11m08.5s
07:10	03°24.4'N	069°19.4'E	00°45.6'N	070°44.6'E	02°04.1'N	070°01.9'E	66	332	11m06.0s
07:15	04°04.3'N	070°22.7'E	01°27.2'N	071°49.6'E	02°44.9'N	071°06.0'E	66	330	11m02.5s
07:20	04°46.7'N	071°25.9'E	02°11.2'N	072°54.4'E	03°28.1'N	072°10.0'E	65	329	10m58.2s
07:25	05°31.7'N	072°29.5'E	02°57.7'N	073°59.3'E	04°13.8'N	073°14.2'E	65	327	10m53.2s
07:30	06°19.4'N	073°33.6'E	03°46.6'N	075°04.7'E	05°02.1'N	074°18.9'E	63	326	10m47.4s
07:35	07°09.9'N	074°38.6'E	04°38.1'N	076°10.9'E	05°53.1'N	075°24.5'E	62	325	10m40.9s
07:40	08°03.2'N	075°45.0'E	05°32.3'N	077°18.3'E	06°46.9'N	076°31.4'E	61	324	10m33.8s
07:45	08°59.6'N	076°53.2'E	06°29.2'N	078°27.3'E	07°43.5'N	077°40.0'E	59	323	10m26.1s
07:50	09°59.3'N	078°03.6'E	07°29.2'N	079°38.4'E	08°43.3'N	078°50.7'E	57	323	10m17.8s
07:55	11°02.4'N	079°17.0'E	08°32.3'N	080°52.2'E	09°46.4'N	080°04.3'E	55	323	10m09.1s
08:00	12°09.3'N	080°33.9'E	09°38.9'N	082°09.4'E	10°53.2'N	081°21.3'E	53	323	09m59.9s
08:05	13°20.4'N	081°55.2'E	10°49.3'N	083°30.8'E	12°03.9'N	082°42.6'E	51	324	09m50.2s
08:10	14°36.2'N	083°22.0'E	12°03.9'N	084°57.4'E	13°19.1'N	084°09.3'E	48	325	09m40.1s
08:15	15°57.4'N	084°55.7'E	13°23.2'N	086°30.4'E	14°39.3'N	085°42.6'E	45	326	09m29.5s
08:20	17°24.7'N	086°38.2'E	14°48.1'N	088°11.6'E	16°05.3'N	087°24.3'E	43	327	09m18.4s
08:25	18°59.4'N	088°31.9'E	16°19.5'N	090°03.3'E	17°38.3'N	089°16.9'E	39	329	09m06.7s
08:30	20°43.2'N	090°40.6'E	17°58.9'N	092°08.6'E	19°19.7'N	091°23.7'E	36	332	08m54.5s
08:35	22°38.9'N	093°10.1'E	19°48.4'N	094°32.7'E	21°12.1'N	093°50.2'E	32	335	08m41.4s
08:40	24°50.8'N	096°10.2'E	21°51.7'N	097°23.6'E	23°19.4'N	096°45.2'E	28	339	08m27.3s
08:45	27°28.1'N	100°00.9'E	24°15.2'N	100°56.9'E	25°49.1'N	100°26.1'E	23	344	08m11.5s
08:50	30°54.8'N	105°37.0'E	27°14.3'N	105°49.4'E	29°00.1'N	105°36.9'E	17	351	07m52.7s
08:55	–	–	31°58.7'N	114°52.1'E	34°49.6'N	116°59.7'E	4	367	07m21.4s
Limits	38°24.0'N	120°53.7'E	35°16.0'N	122°26.7'E	36°49.6'N	121°40.9'E	0	373	07m11.5s

TABLE 2.4

PHYSICAL EPHEMERIS OF THE ANTUMBRAL SHADOW
ANNULAR SOLAR ECLIPSE OF 2010 JANUARY 15

$\Delta T = 66.0$ s

Universal Time	Central Line Latitude	Central Line Longitude	Diameter Ratio	Eclipse Obscur.	Sun Alt °	Sun Azm °	Path Width km	Major Axis km	Minor Axis km	Umbra Veloc. km/s	Central Durat.
05:17.6	06°58.7'N	015°38.6'E	0.9061	0.8210	0.0	111.3	370.7	–	361.0	–	07m09.4s
05:20	03°16.3'N	025°32.1'E	0.9088	0.8260	11.1	112.3	360.4	1817.1	349.5	3.820	07m41.3s
05:25	00°58.9'N	032°22.3'E	0.9108	0.8296	19.4	112.9	354.2	1025.7	341.1	2.017	08m09.8s
05:30	00°18.8'S	036°48.0'E	0.9121	0.8320	25.2	113.3	350.8	789.6	335.6	1.476	08m31.7s
05:35	01°11.1'S	040°15.3'E	0.9132	0.8339	29.8	113.8	348.8	666.5	331.3	1.193	08m50.8s
05:40	01°48.0'S	043°08.9'E	0.9140	0.8354	33.8	114.4	347.4	588.6	327.8	1.013	09m08.1s
05:45	02°14.0'S	045°39.9'E	0.9147	0.8367	37.4	115.1	346.6	534.2	324.8	0.888	09m24.0s
05:50	02°31.6'S	047°54.5'E	0.9154	0.8379	40.7	116.0	346.1	493.9	322.2	0.794	09m38.7s
05:55	02°42.5'S	049°56.6'E	0.9159	0.8389	43.7	117.0	345.8	462.7	319.9	0.723	09m52.3s
06:00	02°47.8'S	051°48.6'E	0.9164	0.8398	46.5	118.2	345.7	438.0	317.8	0.666	10m04.9s
06:05	02°48.2'S	053°32.5'E	0.9168	0.8406	49.1	119.7	345.6	417.9	316.1	0.620	10m16.4s
06:10	02°44.2'S	055°09.7'E	0.9172	0.8413	51.6	121.4	345.5	401.4	314.5	0.583	10m26.9s
06:15	02°36.4'S	056°41.2'E	0.9176	0.8419	53.9	123.3	345.2	387.8	313.1	0.553	10m36.3s
06:20	02°25.1'S	058°07.9'E	0.9178	0.8424	56.0	125.6	344.8	376.4	311.9	0.528	10m44.5s
06:25	02°10.5'S	059°30.4'E	0.9181	0.8429	57.9	128.2	344.3	366.8	310.8	0.508	10m51.7s
06:30	01°52.8'S	060°49.3'E	0.9183	0.8433	59.7	131.2	343.6	358.9	309.9	0.492	10m57.7s
06:35	01°32.3'S	062°05.2'E	0.9185	0.8437	61.3	134.6	342.6	352.4	309.1	0.480	11m02.5s
06:40	01°09.0'S	063°18.4'E	0.9187	0.8440	62.7	138.4	341.5	347.0	308.4	0.470	11m06.2s
06:45	00°43.1'S	064°29.4'E	0.9188	0.8442	63.9	142.7	340.2	342.7	307.9	0.464	11m08.8s
06:50	00°14.6'S	065°38.4'E	0.9189	0.8444	64.9	147.3	338.7	339.5	307.5	0.460	11m10.3s
06:55	00°16.4'N	066°45.9'E	0.9190	0.8445	65.7	152.4	337.1	337.1	307.2	0.458	11m10.8s
07:00	00°49.9'N	067°52.1'E	0.9190	0.8446	66.2	157.7	335.4	335.6	307.0	0.458	11m10.1s
07:05	01°25.8'N	068°57.3'E	0.9190	0.8446	66.4	163.2	333.7	335.0	307.0	0.461	11m08.5s
07:10	02°04.1'N	070°01.9'E	0.9190	0.8446	66.3	168.8	332.0	335.2	307.0	0.466	11m06.0s
07:15	02°44.9'N	071°06.0'E	0.9190	0.8445	66.0	174.3	330.3	336.2	307.2	0.472	11m02.5s
07:20	03°28.1'N	072°10.0'E	0.9189	0.8444	65.4	179.6	328.7	338.1	307.5	0.480	10m58.2s
07:25	04°13.8'N	073°14.2'E	0.9188	0.8442	64.6	184.7	327.3	340.9	307.9	0.491	10m53.2s
07:30	05°02.1'N	074°18.9'E	0.9187	0.8440	63.5	189.4	326.0	344.6	308.4	0.504	10m47.4s
07:35	05°53.1'N	075°24.5'E	0.9185	0.8437	62.2	193.7	324.9	349.4	309.1	0.519	10m40.9s
07:40	06°46.9'N	076°31.4'E	0.9183	0.8433	60.7	197.7	324.1	355.3	309.9	0.536	10m33.8s
07:45	07°43.5'N	077°40.0'E	0.9181	0.8429	59.0	201.3	323.4	362.4	310.8	0.556	10m26.1s
07:50	08°43.3'N	078°50.7'E	0.9179	0.8425	57.1	204.6	323.1	371.1	311.8	0.580	10m17.8s
07:55	09°46.4'N	080°04.3'E	0.9176	0.8419	55.1	207.6	323.0	381.4	313.0	0.607	10m09.1s
08:00	10°53.2'N	081°21.3'E	0.9172	0.8413	52.9	210.3	323.2	393.7	314.4	0.639	09m59.9s
08:05	12°03.9'N	082°42.6'E	0.9169	0.8406	50.6	212.8	323.7	408.6	315.9	0.676	09m50.2s
08:10	13°19.1'N	084°09.3'E	0.9164	0.8399	48.1	215.1	324.5	426.5	317.7	0.720	09m40.1s
08:15	14°39.3'N	085°42.6'E	0.9160	0.8390	45.4	217.3	325.7	448.4	319.6	0.773	09m29.5s
08:20	16°05.3'N	087°24.3'E	0.9155	0.8380	42.5	219.4	327.3	475.6	321.8	0.838	09m18.4s
08:25	17°38.3'N	089°16.9'E	0.9149	0.8370	39.4	221.4	329.3	510.2	324.3	0.919	09m06.7s
08:30	19°19.7'N	091°23.7'E	0.9142	0.8357	36.0	223.4	331.8	555.6	327.1	1.025	08m54.5s
08:35	21°12.1'N	093°50.2'E	0.9134	0.8343	32.2	225.4	334.9	618.1	330.4	1.170	08m41.4s
08:40	23°19.4'N	096°45.2'E	0.9125	0.8326	28.0	227.7	338.8	710.6	334.3	1.382	08m27.3s
08:45	25°49.1'N	100°26.1'E	0.9113	0.8305	23.0	230.2	343.8	866.2	339.0	1.738	08m11.5s
08:50	29°00.1'N	105°36.9'E	0.9098	0.8277	16.5	233.6	351.0	1209.8	345.4	2.519	07m52.7s
08:55	34°49.6'N	116°59.7'E	0.9068	0.8223	4.4	240.5	366.6	4618.8	358.0	10.219	07m21.4s
08:55.4	36°49.6'N	121°40.9'E	0.9058	0.8204	0.0	243.2	372.7	–	362.6	–	07m11.5s

Annular Solar Eclipse of 2010 January 15

TABLE 2.5

LOCAL CIRCUMSTANCES ON THE CENTRAL LINE
ANNULAR SOLAR ECLIPSE OF 2010 JANUARY 15

$\Delta T = 66.0$ s

Central Line Maximum Eclipse			First Contact				Second Contact			Third Contact			Fourth Contact			
U.T.	Durat.	Alt	U.T.	P	V	Alt	U.T.	P	V	U.T.	P	V	U.T.	P	V	Alt
05:18	07m09.5s	1	–	–	–	–	–	–	–	05:21:10	84	167	06:37:27	83	158	18
05:20	07m41.3s	11	–	–	–	–	05:16:10	264	346	05:23:51	83	165	06:50:06	82	154	32
05:25	08m09.8s	19	04:05:54	264	352	1	05:20:56	262	344	05:29:06	82	163	07:04:02	80	148	42
05:30	08m31.7s	25	04:06:42	263	351	6	05:25:45	261	342	05:34:17	81	161	07:15:34	78	142	49
05:35	08m50.8s	30	04:08:00	262	350	10	05:30:35	260	340	05:39:26	80	158	07:25:55	76	136	54
05:40	09m08.1s	34	04:09:37	262	349	13	05:35:27	259	337	05:44:35	79	156	07:35:23	74	129	59
05:45	09m24.0s	37	04:11:26	261	347	16	05:40:19	258	335	05:49:43	78	153	07:44:08	73	122	62
05:50	09m38.7s	41	04:13:25	260	346	19	05:45:12	257	332	05:54:50	77	150	07:52:15	71	114	66
05:55	09m52.3s	44	04:15:31	260	344	21	05:50:05	256	329	05:59:57	75	147	07:59:47	69	104	68
06:00	10m04.9s	47	04:17:45	259	343	23	05:54:59	255	326	06:05:03	74	144	08:06:47	68	94	70
06:05	10m16.4s	49	04:20:05	258	341	25	05:59:53	254	323	06:10:09	73	141	08:13:19	66	83	71
06:10	10m26.9s	52	04:22:31	257	339	27	06:04:47	252	320	06:15:14	72	137	08:19:26	65	72	71
06:15	10m36.3s	54	04:25:03	256	337	29	06:09:43	251	316	06:20:19	71	133	08:25:10	64	62	71
06:20	10m44.5s	56	04:27:41	256	335	31	06:14:38	250	312	06:25:23	69	129	08:30:34	63	52	71
06:25	10m51.7s	58	04:30:26	255	333	33	06:19:35	249	308	06:30:26	68	124	08:35:40	62	44	70
06:30	10m57.7s	60	04:33:18	254	331	35	06:24:32	248	303	06:35:29	67	119	08:40:30	61	36	69
06:35	11m02.5s	61	04:36:16	253	328	36	06:29:29	247	298	06:40:32	66	114	08:45:06	60	30	68
06:40	11m06.2s	63	04:39:22	252	326	38	06:34:27	245	293	06:45:33	65	108	08:49:29	59	24	66
06:45	11m08.8s	64	04:42:36	251	323	40	06:39:26	244	287	06:50:34	64	102	08:53:42	58	20	65
06:50	11m10.3s	65	04:45:59	250	321	41	06:44:25	243	281	06:55:35	63	95	08:57:44	58	16	63
06:55	11m10.8s	66	04:49:31	249	318	43	06:49:25	242	275	07:00:35	62	88	09:01:39	57	12	61
07:00	11m10.1s	66	04:53:13	248	315	44	06:54:25	241	268	07:05:35	61	82	09:05:26	56	9	59
07:05	11m08.5s	66	04:57:06	247	312	46	06:59:25	240	262	07:10:34	60	75	09:09:06	56	7	58
07:10	11m06.0s	66	05:01:09	246	308	47	07:04:27	240	255	07:15:32	59	68	09:12:40	56	5	56
07:15	11m02.5s	66	05:05:25	245	304	48	07:09:28	239	248	07:20:31	58	61	09:16:08	55	3	54
07:20	10m58.2s	65	05:09:54	244	301	50	07:14:30	238	242	07:25:28	58	55	09:19:32	55	2	52
07:25	10m53.2s	65	05:14:36	243	297	51	07:19:33	237	236	07:30:26	57	49	09:22:51	55	0	50
07:30	10m47.4s	63	05:19:32	242	292	52	07:24:36	237	230	07:35:23	56	43	09:26:07	55	359	48
07:35	10m40.9s	62	05:24:44	242	287	53	07:29:39	236	224	07:40:20	56	39	09:29:18	54	358	46
07:40	10m33.8s	61	05:30:11	241	282	54	07:34:42	236	219	07:45:16	56	34	09:32:26	54	358	44
07:45	10m26.1s	59	05:35:53	240	277	55	07:39:46	235	215	07:50:12	55	30	09:35:31	54	357	42
07:50	10m17.8s	57	05:41:53	239	272	55	07:44:50	235	211	07:55:08	55	27	09:38:32	54	357	40
07:55	10m09.1s	55	05:48:09	238	266	56	07:49:55	235	208	08:00:04	55	24	09:41:30	55	356	38
08:00	09m59.9s	53	05:54:43	238	260	56	07:54:59	235	204	08:04:59	55	21	09:44:25	55	356	35
08:05	09m50.2s	51	06:01:34	237	253	56	08:00:04	235	202	08:09:54	55	18	09:47:16	55	356	33
08:10	09m40.1s	48	06:08:42	237	247	55	08:05:09	235	199	08:14:49	55	16	09:50:03	55	356	31
08:15	09m29.5s	45	06:16:09	236	241	54	08:10:14	235	197	08:19:44	55	15	09:52:46	56	356	28
08:20	09m18.4s	43	06:23:53	236	234	53	08:15:20	235	195	08:24:38	55	13	09:55:25	56	357	25
08:25	09m06.7s	39	06:31:57	236	228	51	08:20:26	236	194	08:29:33	55	12	09:57:57	57	357	23
08:30	08m54.5s	36	06:40:21	236	223	49	08:25:32	236	193	08:34:27	56	11	10:00:22	57	358	19
08:35	08m41.4s	32	06:49:08	236	218	46	08:30:39	236	192	08:39:20	56	10	10:02:38	58	359	16
08:40	08m27.3s	28	06:58:22	237	213	42	08:35:46	237	191	08:44:13	57	10	10:04:41	59	360	12
08:45	08m11.5s	23	07:08:14	237	208	37	08:40:54	238	191	08:49:05	58	10	10:06:23	60	1	8
08:50	07m52.7s	17	07:19:10	238	205	31	08:46:03	239	191	08:53:56	60	10	10:07:25	61	3	2
08:55	07m21.4s	5	07:34:09	241	201	18	08:51:19	242	192	08:58:40	62	12	–	–	–	–

TABLE 2.6

TOPOCENTRIC DATA AND PATH CORRECTIONS DUE TO LUNAR LIMB PROFILE
ANNULAR SOLAR ECLIPSE OF 2010 JANUARY 15

$\Delta T = 66.0$ s

Universal Time	Moon Topo H.P. "	Moon Topo S.D. "	Moon Rel. Ang.V "/s	Topo Lib. Long °	Sun Alt. °	Sun Az. °	Path Az. °	North Limit P.A. °	North Limit Int. '	North Limit Ext. '	South Limit Int. '	South Limit Ext. '	Central Durat. Corr. s
05:20	3256.0	886.6	0.386	2.42	11.1	112.3	109.6	173.5	-2.9	2.9	0.9	-4.2	-8.5
05:25	3263.1	888.6	0.355	2.38	19.4	112.9	107.3	172.4	-2.4	3.3	0.9	-3.6	-10.1
05:30	3267.9	889.9	0.335	2.34	25.2	113.3	105.2	171.3	-2.3	3.2	0.8	-3.3	-11.4
05:35	3271.6	890.9	0.319	2.30	29.8	113.8	103.0	170.2	-2.3	3.5	0.7	-3.4	-12.6
05:40	3274.6	891.7	0.306	2.26	33.8	114.4	100.9	169.1	-1.8	3.0	0.4	-2.9	-13.5
05:45	3277.2	892.4	0.295	2.21	37.4	115.1	98.6	168.0	-1.7	2.5	0.3	-2.9	-14.6
05:50	3279.5	893.0	0.285	2.17	40.7	116.0	96.3	166.9	-2.4	2.1	0.7	-3.3	-15.5
05:55	3281.5	893.5	0.277	2.13	43.7	117.0	93.9	165.7	-2.8	1.7	0.8	-2.8	-16.3
06:00	3283.3	894.0	0.270	2.09	46.5	118.2	91.4	164.5	-2.8	1.6	0.8	-2.8	-17.4
06:05	3284.9	894.4	0.263	2.04	49.1	119.7	88.9	163.3	-2.4	1.9	0.6	-2.8	-17.8
06:10	3286.2	894.8	0.258	2.00	51.6	121.4	86.4	162.2	-1.5	2.2	0.5	-3.0	-18.0
06:15	3287.5	895.1	0.253	1.96	53.9	123.3	83.8	161.0	-1.5	2.8	0.9	-4.1	-17.8
06:20	3288.5	895.4	0.249	1.92	56.0	125.6	81.2	159.8	-2.2	2.9	1.0	-4.5	-17.7
06:25	3289.5	895.7	0.245	1.87	57.9	128.2	78.7	158.6	-2.7	2.0	0.6	-4.9	-17.4
06:30	3290.3	895.9	0.242	1.83	59.7	131.2	76.1	157.4	-2.9	2.5	0.7	-5.2	-17.1
06:35	3291.0	896.1	0.240	1.79	61.3	134.6	73.6	156.3	-3.0	3.0	1.1	-4.9	-16.7
06:40	3291.6	896.2	0.238	1.75	62.7	138.4	71.1	155.2	-2.7	3.1	1.0	-4.4	-16.2
06:45	3292.0	896.4	0.237	1.70	63.9	142.7	68.7	154.1	-2.5	3.6	1.0	-4.2	-16.0
06:50	3292.4	896.5	0.236	1.66	64.9	147.3	66.4	153.0	-3.4	4.1	1.4	-4.7	-15.9
06:55	3292.6	896.5	0.236	1.62	65.7	152.4	64.2	152.0	-3.9	4.6	1.5	-4.2	-15.7
07:00	3292.8	896.6	0.236	1.58	66.2	157.7	62.2	151.1	-4.1	3.5	1.2	-3.9	-15.4
07:05	3292.8	896.6	0.236	1.53	66.4	163.2	60.2	150.2	-4.6	4.0	1.1	-4.2	-15.0
07:10	3292.8	896.6	0.237	1.49	66.3	168.8	58.4	149.3	-4.7	3.7	1.1	-4.5	-14.3
07:15	3292.6	896.5	0.239	1.45	66.0	174.3	56.7	148.5	-4.7	4.1	1.2	-4.8	-13.8
07:20	3292.4	896.5	0.240	1.41	65.4	179.6	55.2	147.8	-4.4	4.5	1.3	-4.5	-13.3
07:25	3292.0	896.4	0.243	1.36	64.6	184.7	53.8	147.2	-4.1	4.9	1.2	-4.3	-12.8
07:30	3291.5	896.2	0.245	1.32	63.5	189.4	52.6	146.6	-3.7	5.2	1.2	-4.2	-12.4
07:35	3291.0	896.1	0.248	1.28	62.2	193.7	51.5	146.1	-3.7	5.7	1.3	-4.1	-12.2
07:40	3290.3	895.9	0.251	1.24	60.7	197.7	50.6	145.7	-4.1	5.6	1.2	-4.0	-12.0
07:45	3289.5	895.7	0.255	1.19	59.0	201.3	49.8	145.3	-4.3	4.9	1.1	-3.9	-11.7
07:50	3288.5	895.4	0.259	1.15	57.1	204.6	49.2	145.0	-4.4	4.5	1.0	-3.7	-11.4
07:55	3287.5	895.2	0.264	1.11	55.1	207.6	48.8	144.8	-4.5	4.6	0.9	-3.8	-11.1
08:00	3286.3	894.8	0.269	1.07	52.9	210.3	48.5	144.7	-4.5	4.7	0.9	-3.9	-10.7
08:05	3284.9	894.5	0.275	1.02	50.6	212.8	48.4	144.6	-4.5	4.8	0.9	-3.9	-10.5
08:10	3283.4	894.1	0.281	0.98	48.1	215.1	48.4	144.7	-4.5	4.7	0.9	-3.9	-10.3
08:15	3281.7	893.6	0.288	0.94	45.4	217.3	48.6	144.8	-4.4	4.7	1.0	-3.9	-10.1
08:20	3279.8	893.1	0.295	0.90	42.5	219.4	48.9	145.1	-4.3	5.3	1.1	-4.0	-9.8
08:25	3277.6	892.5	0.304	0.86	39.4	221.4	49.5	145.4	-4.2	5.5	1.2	-4.2	-9.7
08:30	3275.1	891.8	0.313	0.81	36.0	223.4	50.2	145.8	-3.9	6.0	1.3	-4.4	-9.5
08:35	3272.3	891.1	0.324	0.77	32.2	225.4	51.2	146.4	-3.4	5.7	1.3	-4.7	-9.3
08:40	3268.9	890.2	0.337	0.73	28.0	227.7	52.5	147.2	-3.9	5.2	1.1	-5.0	-9.1
08:45	3264.8	889.0	0.352	0.69	23.0	230.2	54.1	148.2	-4.4	4.6	1.2	-5.5	-9.2
08:50	3259.3	887.6	0.372	0.64	16.5	233.6	56.5	149.5	-4.5	3.9	1.1	-4.7	-9.4
08:55	3248.7	884.7	0.412	0.60	4.4	240.5	61.3	152.3	-3.8	4.6	1.3	-4.9	-9.0

ANNULAR SOLAR ECLIPSE OF 2010 JANUARY 15

TABLE 2.7
LOCAL CIRCUMSTANCES FOR AFRICA: ANGOLA TO LIBYA
ANNULAR SOLAR ECLIPSE OF 2010 JANUARY 15

Location Name	Latitude	Longitude	Elev.	First Contact U.T. h m s	P °	V °	Alt °	Second Contact U.T. h m s	P °	V °	Third Contact U.T. h m s	P °	V °	Fourth Contact U.T. h m s	P °	V °	Alt °	Maximum Eclipse U.T. h m s	P °	V °	Alt °	Azm °	Eclip. Mag.	Eclip. Obs.	Umbral Depth	Umbral Durat.
ANGOLA																										
Luanda	08°48'S	013°14'E	59	—				—			—			06:21:43.1	56	148	18	05:14:47.9	355	93	3	111	0.508	0.387		
BENIN																										
Cotonou	06°21'N	002°26'E	—	—				—			—			06:26:55.3	75	157	4	06:07 Rise	—	—	0	111	0.253	0.142		
Porto-Novo	06°29'N	002°37'E	—	—				—			—			06:27:06.8	76	157	4	06:07 Rise	—	—	0	111	0.262	0.150		
BOTSWANA																										
Gaborone	24°45'S	025°55'E	—	04:38:01.5	315	67	11	—			—			06:14:32.5	29	135	33	05:23:58.0	352	102	21	104	0.198	0.100		
BURUNDI																										
Bujumbura	03°23'S	029°22'E	—	04:05:43.7	273	6	0	—			—			06:53:40.7	72	149	39	05:21:02.1	353	80	17	111	0.812	0.732		
CAMEROON																										
Douala	04°03'N	009°42'E	—	—				—			—			06:29:48.1	76	157	12	05:34 Rise	—	—	0	111	0.668	0.565		
Yaoundé	03°52'N	011°31'E	770	—				—			—			06:31:14.1	76	157	14	05:27 Rise	—	—	0	111	0.760	0.671		
CENTRAL AFRICAN REPUBLIC																										
Bangui	04°22'N	018°35'E	387	—				05:15:09.1	321	45	05:19:10.6	28	111	06:39:30.5	81	156	23	05:17:10.1	354	78	4	112	0.907	0.823	0.167	04m02s
Bouar	05°57'N	015°36'E	—	—				05:14:13.0	300	24	05:20:01.2	49	132	06:36:47.3	82	158	19	05:17:07.0	354	78	1	111	0.906	0.821	0.417	05m48s
CHAD																										
Moundou	08°34'N	016°05'E	—	—				05:17:38.4	187	268	05:19:15.2	161	242	06:38:49.7	86	159	18	05:18:27.2	174	255	1	111	0.906	0.821	0.025	01m37s
Ndjamena	12°07'N	015°03'E	295	—				—			—			06:39:33.4	91	161	16	05:26 Rise	—	—	0	112	0.809	0.726		
CONGO																										
Brazzaville	04°16'S	015°17'E	318	—				—			—			06:28:47.8	65	152	20	05:14:45.0	354	88	3	111	0.651	0.545		
Pointe-Noire	04°48'S	011°51'E	50	—				—			—			06:24:22.5	62	151	16	05:14:34.7	355	90	1	111	0.601	0.489		
DEM. REP. CONGO																										
Beni	00°30'N	029°28'E	—	—				05:19:51.1	320	43	05:24:10.2	26	109	06:56:41.7	78	150	38	05:22:00.9	353	76	16	112	0.910	0.828	0.161	04m19s
Bunia	01°34'N	030°11'E	—	—				05:19:05.9	266	348	05:27:05.4	80	161	06:59:00.9	80	150	39	05:23:05.0	353	74	17	113	0.910	0.828	0.944	08m00s
Butembo	00°09'N	029°17'E	—	—				—			—			06:56:03.1	78	150	38	05:21:45.1	353	76	16	112	0.906	0.827		
Isiro	02°47'N	027°37'E	—	—				05:17:25.8	255	337	05:25:10.8	91	172	06:53:59.9	82	152	35	05:21:17.7	173	255	14	113	0.909	0.827	0.861	07m45s
Kananga	05°54'S	022°25'E	—	—				—			—			06:53:35.7	66	150	30	05:16:35.7	354	86	11	110	0.679	0.578		
Kinshasa	04°18'S	015°18'E	—	—				—			—			06:37:46.5	65	152	20	05:14:45.4	354	88	3	111	0.650	0.545		
Kisangani	00°30'S	025°11'E	—	—				—			—			06:47:07.6	76	152	32	05:18:37.6	354	79	12	112	0.851	0.775		
Kolwezi	10°43'S	025°28'E	—	04:11:06.8	287	29	1	—			—			06:38:21.8	58	147	34	05:18:26.1	353	89	16	109	0.577	0.463		
Lubumbashi	11°40'S	027°28'E	—	04:11:24.9	288	29	3	—			—			06:41:13.8	57	146	37	05:19:47.1	353	89	18	109	0.569	0.454		
Mbuji-Mayi	06°09'S	023°38'E	—	—				—			—			06:39:40.8	66	150	31	05:17:08.5	354	86	12	110	0.684	0.584		
DJIBOUTI																										
Djibouti	11°36'N	043°09'E	7	04:22:23.3	240	313	11	—			—			07:42:18.3	94	132	50	05:50:30.2	168	230	30	123	0.616	0.507		
EGYPT																										
Cairo	30°03'N	031°15'E	116	—				—			—			07:04:33.6	122	164	23	05:53:45.1	170	222	11	122	0.289	0.173		
EQUATORIAL GUINEA																										
Malabo, Bioko	03°45'N	008°47'E	—	—				—			—			06:28:50.7	75	156	11	05:38 Rise	—	—	0	111	0.620	0.510		
ERITREA																										
Asmera	15°20'N	038°53'E	2325	04:24:26.1	237	308	6	—			—			07:27:05.1	101	144	42	05:45:31.2	170	231	23	122	0.549	0.433		
ETHIOPIA																										
Addis Ababa	09°02'N	038°42'E	2450	04:14:29.2	247	325	6	—			—			07:25:58.8	92	141	46	05:38:44.9	170	239	25	118	0.701	0.605		
GABON																										
Libreville	00°23'N	009°27'E	35	—				—			—			06:26:43.6	70	154	13	05:30 Rise	—	—	0	111	0.642	0.535		
GHANA																										
Accra	05°33'N	000°13'W	27	—				—			—			06:25:09.1	72	156	2	06:16 Rise	—	—	0	111	0.112	0.043		
KENYA																										
Eldoret	00°31'N	035°17'E	—	04:06:25.4	263	351	4	05:24:12.6	248	328	05:32:21.7	96	175	07:11:40.0	79	145	46	05:28:16.5	172	252	23	113	0.912	0.831	0.756	08m09s
Kisumu	00°06'S	034°45'E	—	04:06:07.7	264	353	4	05:23:18.4	275	356	05:31:27.0	69	149	07:09:44.4	78	145	46	05:27:22.0	352	73	23	113	0.912	0.831	0.779	08m09s
Machakos	01°31'S	037°16'E	—	04:06:38.1	265	354	7	05:27:01.1	302	24	05:33:26.3	40	120	07:16:04.9	76	141	50	05:30:13.6	351	72	26	113	0.912	0.832	0.339	06m25s
Meru	00°03'N	037°39'E	—	04:07:07.0	262	349	7	05:27:21.9	240	320	05:35:24.4	102	181	07:18:26.0	78	141	50	05:31:22.6	171	250	26	114	0.912	0.832	0.642	08m03s
Mombasa	04°03'S	039°40'E	16	04:07:28.5	267	358	10	—			—			07:21:21.8	72	137	54	05:33:02.3	350	73	30	112	0.875	0.804		
Nairobi	01°17'S	036°49'E	1820	04:06:28.6	265	354	6	05:26:11.6	297	18	05:33:05.0	45	126	07:14:22.7	76	142	49	05:29:38.0	351	72	25	113	0.912	0.832	0.412	06m53s
Nakuru	00°17'S	036°04'E	—	04:06:27.5	263	352	5	05:24:48.5	268	348	05:33:13.4	75	155	07:13:24.2	78	143	48	05:29:00.2	351	72	24	113	0.912	0.832	0.894	08m25s
LESOTHO																										
Maseru	29°28'S	027°30'E	—	04:56:04.4	328	83	18	—			—			06:01:21.1	15	127	32	05:27:48.1	352	105	25	101	0.084	0.028		
LIBYA																										
Banghazi	32°07'N	020°04'E	25	—				—			—			06:50:09.3	122	170	12	05:45:13.7	172	227	1	115	0.328	0.208		
Tripoli	32°54'N	013°11'E	22	—				—			—			06:45:58.6	120	171	6	06:12 Rise	—	—	0	115	0.268	0.155		

TABLE 2.8
LOCAL CIRCUMSTANCES FOR AFRICA: MADAGASCAR TO ZIMBABWE
ANNULAR SOLAR ECLIPSE OF 2010 JANUARY 15

Location Name	Latitude	Longitude	Elev.	First Contact U.T. h m s	P °	V °	Alt °	Second Contact U.T. h m s	P °	V °	Third Contact U.T. h m s	P °	V °	Fourth Contact U.T. h m s	P °	V °	Alt °	Maximum Eclipse U.T. h m s	P °	V °	Alt °	Azm °	Eclip. Mag.	Eclip. Obs.	Umbral Depth	Umbral Durat.
			m																							
MADAGASCAR																										
Antananarivo	18°55'S	047°31'E	—	04:23:50.7	288	30	26	—	—	—	—	—	—	07:20:54.8	44	133	67	05:44:53.3	347	83	44	101	0.497	0.377		
Fianarantsoa	21°26'S	047°05'E	—	04:28:19.9	294	37	27	—	—	—	—	—	—	07:13:07.1	40	135	65	05:44:23.3	347	87	44	99	0.423	0.300		
MALAWI																										
Blantyre	15°47'S	035°00'E	—	04:15:13.8	290	33	12	—	—	—	—	—	—	06:52:44.4	52	142	48	05:26:58.0	351	88	28	106	0.517	0.398		
Lilongwe	13°59'S	033°44'E	—	04:12:53.0	288	29	9	—	—	—	—	—	—	06:52:23.3	55	143	46	05:25:17.9	352	87	26	107	0.558	0.442		
MAYOTTE																										
Dzaoudzi	12°47'S	045°17'E	—	04:14:29.4	279	15	20	—	—	—	—	—	—	07:27:23.7	56	131	65	05:41:02.5	348	77	40	107	0.664	0.562		
MOZAMBIQUE																										
Beira	19°49'S	034°52'E	9	04:22:07.5	298	44	14	—	—	—	—	—	—	06:45:03.4	44	140	47	05:28:03.9	351	93	29	104	0.403	0.280		
Maputo	25°58'S	032°35'E	59	04:38:25.6	313	64	18	—	—	—	—	—	—	06:25:05.3	29	135	41	05:28:57.0	351	100	29	101	0.216	0.114		
NAMIBIA																										
Windhoek	22°34'S	017°06'E	1728	04:38:53.9	318	71	3	—	—	—	—	—	—	06:04:07.6	29	135	22	05:19:45.1	353	103	12	108	0.179	0.086		
NIGER																										
Niamey	13°31'N	002°07'E	216	—				06:19:44.1			06:19:44.1			06:32:00.6	86	161	2	06:20 Rise	—	—	0	112	0.156	0.070		
NIGERIA																										
Ibadan	07°17'N	003°30'E	—	—				—	—	—	—	—	—	06:28:09.3	77	158	5	06:04 Rise	—	—	0	111	0.303	0.186		
Kano	12°00'N	008°30'E	—	—				—	—	—	—	—	—	06:34:10.8	87	161	9	05:52 Rise	—	—	0	112	0.521	0.402		
Lagos	06°27'N	003°24'E	—	—				—	—	—	—	—	—	06:27:28.1	76	157	5	06:03 Rise	—	—	0	111	0.306	0.188		
Ogbomosho	08°08'N	004°15'E	3	—				—	—	—	—	—	—	06:29:09.8	79	158	6	06:03 Rise	—	—	0	111	0.337	0.216		
RWANDA																										
Kigali	01°57'S	030°04'E	—	04:05:29.0	270	2	0	—	—	—	—	—	—	06:56:23.8	75	149	39	05:21:53.7	353	78	18	112	0.857	0.783		
SOMALIA																										
Kismaayo	00°22'S	042°32'E	—	04:09:38.6	260	345	12	05:36:51.6	205	282	05:42:09.2	134	210	07:34:28.5	77	131	57	05:39:30.6	169	245	33	115	0.914	0.835	0.187	05m18s
Moqadisho	02°04'N	045°22'E	12	04:13:17.9	254	336	15	—	—	—	—	—	—	07:46:58.4	79	122	60	05:47:00.8	168	239	36	118	0.845	0.771		
SOUTH AFRICA																										
Bloemfontein	29°12'S	026°07'E	—	04:56:07.4	329	84	17	—	—	—	—	—	—	05:55:22.8	15	127	30	05:25:54.1	352	105	23	102	0.080	0.026		
Durban	29°55'S	030°56'E	5	04:55:22.8	326	81	21	—	—	—	—	—	—	06:07:15.8	16	128	36	05:30:11.9	351	104	28	99	0.096	0.034		
Johannesburg	26°15'S	028°00'E	—	04:41:42.8	317	70	14	—	—	—	—	—	—	06:14:27.2	26	134	35	05:25:59.6	352	102	24	103	0.174	0.083		
Pretoria	25°45'S	028°10'E	1369	04:39:50.7	316	68	14	—	—	—	—	—	—	06:16:15.6	27	135	35	05:25:45.9	352	101	24	103	0.189	0.093		
SUDAN																										
Khartoum	15°36'N	032°32'E	390	—				—	—	—	—	—	—	07:09:06.7	102	154	34	05:35:30.8	171	238	16	118	0.585	0.472		
SWAZILAND																										
Mbabane	26°18'S	031°06'E	—	04:40:09.6	315	67	17	—	—	—	—	—	—	06:20:52.4	27	135	39	05:28:01.9	351	101	27	101	0.196	0.099		
TANZANIA																										
Dar-es-Salaam	06°48'S	039°17'E	14	04:07:54.9	272	5	11	—	—	—	—	—	—	07:17:11.9	67	138	54	05:31:51.0	351	76	30	111	0.798	0.717		
Mwanza	02°31'S	032°54'E	—	04:05:37.9	269	1	3	—	—	—	—	—	—	07:02:57.9	74	146	44	05:24:31.7	352	77	21	112	0.866	0.793		
TOGO																										
Lome	06°08'N	001°13'E	22	—				—	—	—	—	—	—	06:26:12.5	74	156	3	06:12 Rise	—	—	0	111	0.187	0.092		
TUNISIA																										
Tunis	36°48'N	010°11'E	66	—				—	—	—	—	—	—	06:46:12.9	123	172	2	06:33 Rise	—	—	0	116	0.111	0.042		
UGANDA																										
Jinja	00°26'N	033°12'E	—	04:05:54.6	264	353	2	05:21:40.9	272	354	05:29:46.6	72	153	07:05:51.4	79	147	43	05:25:43.1	352	73	21	113	0.911	0.830	0.823	08m06s
Kampala	00°19'N	032°25'E	1312	04:05:44.0	265	354	2	05:21:02.1	285	7	05:28:33.2	59	141	07:03:42.6	79	148	42	05:24:47.1	352	74	20	113	0.911	0.830	0.612	07m31s
Masaka	00°20'S	031°44'E	—	04:05:34.7	266	356	1	05:22:10.2	327	50	05:25:36.0	17	100	07:01:34.3	77	148	41	05:23:53.5	353	75	19	112	0.911	0.829	0.094	03m26s
Mbale	01°05'N	034°10'E	—	04:06:17.5	262	350	3	05:23:16.4	239	320	05:30:56.4	105	184	07:08:53.0	80	146	44	05:27:05.9	172	252	22	113	0.911	0.830	0.612	07m40s
VENDA																										
Thohoyandou	23°00'S	030°29'E	—	04:30:29.5	307	57	13	—	—	—	—	—	—	06:28:17.7	35	138	40	05:25:47.2	352	98	25	103	0.282	0.168		
ZAMBIA																										
Kitwe	12°49'S	028°13'E	—	04:12:29.5	289	31	4	—	—	—	—	—	—	06:41:20.4	56	145	38	05:20:30.2	353	89	19	108	0.544	0.427		
Lusaka	15°25'S	028°17'E	1277	04:15:56.6	294	38	6	—	—	—	—	—	—	06:37:49.0	51	114	38	05:21:10.7	353	92	20	107	0.473	0.351		
Ndola	12°58'S	028°38'E	—	04:12:33.0	289	31	4	—	—	—	—	—	—	06:42:01.7	55	145	39	05:20:49.3	353	89	20	108	0.544	0.427		
ZIMBABWE																										
Bulawayo	20°09'S	028°36'E	1343	04:24:25.0	303	51	9	—	—	—	—	—	—	06:30:23.7	41	141	38	05:23:07.7	352	96	22	105	0.346	0.225		
Harare	17°50'S	031°03'E	1472	04:19:03.3	296	42	10	—	—	—	—	—	—	06:39:54.0	47	142	42	05:23:58.0	352	93	24	106	0.429	0.306		

ANNULAR SOLAR ECLIPSE OF 2010 JANUARY 15

TABLE 2.9
LOCAL CIRCUMSTANCES FOR EUROPE
ANNULAR SOLAR ECLIPSE OF 2010 JANUARY 15

Location Name	Latitude	Longitude	Elev.	First Contact U.T. h m s	P °	V °	Alt °	Second Contact U.T. h m s	P °	V °	Third Contact U.T. h m s	P °	V °	Fourth Contact U.T. h m s	P °	V °	Alt °	Maximum Eclipse U.T. h m s	P °	V °	Alt °	Azm °	Eclip. Mag.	Eclip. Obs.	Umbral Depth	Umbral Durat.
			m																							
ALBANIA																										
Tiranë	41°20'N	019°50'E	7	—	—	—	—	—	—	—	—	—	—	06:50:35.6	134	175	7	06:06 Rise	—	—	0	118	0.171	0.080		
AUSTRIA																										
Vienna	48°13'N	016°20'E	202	—	—	—	—	—	—	—	—	—	—	06:49:59.4	142	179	1	06:43 Rise	—	—	0	122	0.035	0.008		
BOSNIA & HERZEGOWINA																										
Sarajevo	43°52'N	018°25'E	—	—	—	—	—	—	—	—	—	—	—	06:50:06.2	137	177	4	06:19 Rise	—	—	0	119	0.129	0.053		
BULGARIA																										
Sofija	42°41'N	023°19'E	550	—	—	—	—	—	—	—	—	—	—	06:52:22.2	138	176	8	06:07:33.3	169	212	2	121	0.133	0.055		
CROATIA																										
Zagreb	45°48'N	015°58'E	—	—	—	—	—	—	—	—	—	—	—	06:49:41.8	139	178	2	06:35 Rise	—	—	0	120	0.072	0.022		
CZECH REPUBLIC																										
Ostrava	49°50'N	018°17'E	—	—	—	—	—	—	—	—	—	—	—	06:50:06.7	146	180	1	06:41 Rise	—	—	0	123	0.035	0.007		
GREECE																										
Athens	37°58'N	023°43'E	107	—	—	—	—	—	—	—	—	—	—	06:53:16.8	131	173	11	05:58:41.2	170	218	3	119	0.200	0.101		
HUNGARY																										
Budapest	47°30'N	019°05'E	120	—	—	—	—	—	—	—	—	—	—	06:50:17.0	143	179	3	06:29 Rise	—	—	0	122	0.078	0.025		
ITALY																										
Naples	40°51'N	014°17'E	25	—	—	—	—	—	—	—	—	—	—	06:48:26.8	131	175	3	06:27 Rise	—	—	0	118	0.133	0.056		
Rome	41°54'N	012°29'E	115	—	—	—	—	—	—	—	—	—	—	06:48:17.8	131	175	2	06:37 Rise	—	—	0	118	0.075	0.024		
MACEDONIA																										
Skopje	41°59'N	021°26'E	240	—	—	—	—	—	—	—	—	—	—	06:51:23.4	136	176	7	06:04:17.4	170	214	0	119	0.153	0.069		
MOLDOVA																										
Kisin'ov	47°00'N	028°50'E	—	05:53:36.6	188	226	1	—	—	—	—	—	—	06:54:12.5	147	179	8	06:23:17.1	168	203	5	128	0.057	0.016		
POLAND																										
Krakow	50°03'N	019°58'E	220	—	—	—	—	—	—	—	—	—	—	06:50:06.9	147	181	2	06:35 Rise	—	—	0	123	0.047	0.012		
Warsaw	52°15'N	021°00'E	90	—	—	—	—	—	—	—	—	—	—	06:49:28.7	151	183	1	06:41 Rise	—	—	0	125	0.023	0.004		
ROMANIA																										
Bucharest	44°26'N	026°06'E	82	—	—	—	—	—	—	—	—	—	—	06:53:42.7	142	177	9	06:14:19.5	169	208	3	124	0.096	0.035		
RUSSIA																										
Barnaul	53°22'N	083°45'E	—	07:14:50.2	195	189	15	—	—	—	—	—	—	09:12:37.6	114	91	8	08:15:27.6	154	139	12	204	0.227	0.122		
Chelyabinsk	55°10'N	061°24'E	—	07:09:25.3	175	183	13	—	—	—	—	—	—	08:05:41.2	141	141	14	07:37:37.4	158	162	14	174	0.040	0.009		
Irkutsk	52°16'N	104°20'E	467	07:25:17.1	211	191	11	—	—	—	—	—	—	—	—	—	—	08:36:10.0	154	126	4	227	0.439	0.316		
Kemerovo	55°20'N	086°05'E	—	07:17:43.7	195	188	13	—	—	—	—	—	—	09:12:08.1	114	92	6	08:16:31.5	155	139	10	206	0.225	0.120		
Krasnojarsk	56°01'N	092°50'E	152	07:21:06.2	200	188	11	—	—	—	—	—	—	09:20:51.2	110	84	2	08:22:52.5	155	135	7	214	0.275	0.161		
Novokuzneck	53°45'N	087°06'E	—	07:16:56.0	197	189	14	—	—	—	—	—	—	09:17:44.1	111	87	6	08:19:11.2	154	137	11	208	0.253	0.143		
Novosibirsk	55°02'N	082°55'E	—	07:16:03.3	193	187	13	—	—	—	—	—	—	09:06:58.0	117	96	8	08:12:55.3	155	141	11	203	0.200	0.102		
Omsk	55°00'N	073°24'E	85	07:12:04.1	186	186	14	—	—	—	—	—	—	08:44:46.5	126	113	12	07:59:06.8	156	149	13	190	0.122	0.049		
Tomsk	56°30'N	084°58'E	—	07:18:10.7	194	187	12	—	—	—	—	—	—	09:07:15.3	117	96	6	08:14:04.0	155	141	9	205	0.202	0.103		
Ufa	54°44'N	055°56'E	174	07:08:59.0	170	181	12	—	—	—	—	—	—	07:44:20.4	149	155	14	07:26:40.6	159	168	13	166	0.015	0.002		
Vladivostok	43°10'N	131°56'E	29	07:40:17.5	240	199	3	—	—	—	—	—	—	—	—	—	—	08:00 Set	—	—	0	241	0.250	0.140		
Volgograd	48°44'N	044°25'E	—	06:35:30.3	175	200	12	—	—	—	—	—	—	07:14:34.3	151	171	15	06:54:57.6	163	185	14	147	0.018	0.003		
SERBIA AND MONTENEGRO																										
Beograd	44°50'N	020°30'E	138	—	—	—	—	—	—	—	—	—	—	06:50:48.0	140	178	5	06:14 Rise	—	—	0	120	0.119	0.047		
SLOVAKIA																										
Bratislava	48°09'N	017°07'E	—	—	—	—	—	—	—	—	—	—	—	06:50:02.9	143	179	1	06:39 Rise	—	—	0	122	0.048	0.012		
SLOVENIA																										
Ljubljana	46°03'N	014°31'E	—	—	—	—	—	—	—	—	—	—	—	06:49:33.1	138	177	1	06:42 Rise	—	—	0	121	0.041	0.010		
UKRAINE																										
Kiev	50°26'N	030°31'E	—	06:14:32.8	179	212	2	—	—	—	—	—	—	06:51:29.6	154	183	7	06:32:53.8	167	198	4	131	0.020	0.003		
Odessa	46°28'N	030°44'E	65	05:54:59.7	188	226	2	—	—	—	—	—	—	06:56:02.1	147	178	10	06:24:53.2	167	202	6	129	0.055	0.015		

TABLE 2.10
LOCAL CIRCUMSTANCES FOR JAPAN

Location Name	Latitude	Longitude	Elev.	First Contact U.T. h m s	P °	V °	Alt °	Second Contact U.T. h m s	P °	V °	Third Contact U.T. h m s	P °	V °	Fourth Contact U.T. h m s	P °	V °	Alt °	Maximum Eclipse U.T. h m s	P °	V °	Alt °	Azm °	Eclip. Mag.	Eclip. Obs.	Umbral Depth	Umbral Durat.
			m																								
JAPAN																										
Fukuoka	33°35'N	130°24'E	—	07:45:30.8	253	204	8	—	—	—	—	—	—	—	—	—	—	08:31 Set	—	—	0	245	0.560	0.444		
Hiroshima	34°24'N	132°27'E	—	07:46:09.3	253	204	6	—	—	—	—	—	—	—	—	—	—	08:21 Set	—	—	0	245	0.437	0.313		
Kitakyushu	33°53'N	130°50'E	—	07:45:34.8	253	204	7	—	—	—	—	—	—	—	—	—	—	08:29 Set	—	—	0	245	0.533	0.414		
Kobe	34°41'N	135°10'E	—	07:47:18.9	255	204	4	—	—	—	—	—	—	—	—	—	—	08:18 Set	—	—	0	244	0.282	0.167		
Kyoto	35°00'N	135°45'E	—	07:47:19.9	255	204	3	—	—	—	—	—	—	—	—	—	—	08:07 Set	—	—	0	244	0.244	0.135		
Osaka	34°40'N	135°30'E	15	07:47:28.8	255	204	3	—	—	—	—	—	—	—	—	—	—	08:08 Set	—	—	0	244	0.263	0.151		

TABLE 2.11
LOCAL CIRCUMSTANCES FOR THE MIDDLE EAST
ANNULAR SOLAR ECLIPSE OF 2010 JANUARY 15

Location Name	Latitude	Longitude	Elev.	First Contact U.T. h m s	P °	V °	Alt °	Second Contact U.T. h m s	P °	V °	Third Contact U.T. h m s	P °	V °	Fourth Contact U.T. h m s	P °	V °	Alt °	Maximum Eclipse U.T. h m s	P °	V °	Alt °	Azm °	Eclip. Mag.	Eclip. Obs.	Umbral Depth	Umbral Durat.
			m																							
ARMENIA																										
Jerevan	40°11'N	044°30'E	—	05:54:55.2	191	228	14	—			—			07:28:46.8	136	157	24	06:40:26.6	164	193	19	143	0.097	0.035		
AZERBAIJAN																										
Baku	40°23'N	049°51'E	—	06:06:42.1	190	221	18	—			—			07:47:06.3	132	146	27	06:55:48.3	161	184	23	151	0.103	0.038		
BAHRAIN																										
Al-Manamah	26°13'N	050°35'E	—	05:12:54.2	212	263	20	—			—			08:07:36.9	111	122	42	06:34:46.9	162	198	33	143	0.304	0.187		
GEORGIA																										
Tbilisi	41°43'N	044°49'E	—	06:02:23.4	188	223	14	—			—			07:27:44.1	138	159	23	06:44:00.4	163	191	19	144	0.081	0.026		
IRAN																										
Esfahan	32°40'N	051°38'E	1597	05:39:09.2	202	242	21	—			—			08:05:07.9	119	129	35	06:49:21.6	161	188	30	149	0.205	0.105		
Qom	34°39'N	050°54'E	—	05:45:20.1	199	237	20	—			—			07:59:29.3	123	134	33	06:50:04.7	161	188	28	149	0.175	0.084		
Shiraz	29°37'N	052°33'E	—	05:29:33.1	206	250	22	—			—			08:12:51.3	114	121	39	06:47:39.6	160	189	33	149	0.254	0.144		
Tehran	35°40'N	051°26'E	1200	05:50:36.0	197	233	20	—			—			08:00:17.1	124	134	32	06:53:28.4	160	186	28	151	0.163	0.075		
IRAQ																										
Basra	30°30'N	047°47'E	—	05:22:20.5	206	253	17	—			—			07:51:38.7	119	137	36	06:32:42.8	163	198	28	141	0.230	0.125		
Mosul	36°20'N	043°08'E	223	05:35:35.3	199	241	12	—			—			07:29:00.3	131	155	27	06:29:44.8	165	200	20	139	0.146	0.064		
Baghdad	33°21'N	044°25'E	34	05:26:13.3	203	249	14	—			—			07:36:13.3	125	149	31	06:27:42.7	164	201	23	138	0.187	0.092		
ISRAEL																										
Jerusalem	31°46'N	035°14'E	809	05:03:35.2	211	265	4	—			—			07:11:49.2	125	162	25	06:03:22.8	168	216	15	127	0.237	0.130		
JORDAN																										
'Amman	31°57'N	035°56'E	776	05:05:18.9	211	264	5	—			—			07:13:17.2	125	161	26	06:05:02.3	168	215	15	128	0.231	0.125		
KUWAIT																										
Kuwait City	29°20'N	047°59'E	5	05:18:26.0	208	257	17	—			—			07:53:37.7	117	135	37	06:31:18.2	163	199	29	141	0.249	0.140		
LEBANON																										
Beirut	33°53'N	035°30'E	—	05:11:29.9	207	259	5	—			—			07:11:18.6	128	163	24	06:07:47.9	168	213	14	128	0.202	0.103		
OMAN																										
Masqat	23°37'N	058°35'E	—	05:22:39.2	215	260	29	—			—			08:45:23.6	98	88	45	07:01:46.4	156	179	42	157	0.386	0.264		
SAUDI ARABIA																										
Mecca	21°27'N	039°49'E	—	04:39:09.7	226	290	7	—			—			07:28:52.3	110	146	38	05:55:39.1	168	223	23	126	0.411	0.288		
Riyadh	24°38'N	046°43'E	591	04:59:39.5	217	272	15	—			—			07:52:26.8	111	133	41	06:19:04.4	165	208	29	136	0.330	0.210		
SYRIA																										
Damascus	33°30'N	036°18'E	720	05:11:26.5	208	259	5	—			—			07:13:14.6	128	162	25	06:08:37.1	168	212	15	129	0.204	0.105		
TURKEY																										
Ankara	39°56'N	032°52'E	861	05:30:26.7	198	243	3	—			—			07:03:00.1	137	171	17	06:14:50.1	168	208	10	129	0.124	0.050		
Bursa	40°11'N	029°04'E	—	—				—			—			06:57:58.6	137	173	14	06:09:29.6	169	211	7	125	0.137	0.058		
Istanbul	41°01'N	028°58'E	18	—				—			—			06:57:31.6	138	174	13	06:11:02.0	169	210	6	125	0.126	0.052		
Izmir	38°25'N	027°09'E	28	—				—			—			06:56:27.5	133	172	14	06:03:27.1	169	215	5	122	0.173	0.082		
UNITED ARAB EMIRATES																										
Abu Dhabi	24°28'N	054°22'E	—	05:15:24.8	214	263	24	—			—			08:26:04.9	104	106	44	06:46:05.4	159	190	38	148	0.346	0.225		
YEMEN																										
Sana	15°23'N	044°12'E	—	04:30:50.2	233	302	12	—			—			07:46:17.3	99	131	48	05:57:41.7	167	224	30	126	0.524	0.406		

TABLE 2.12
LOCAL CIRCUMSTANCES FOR CENTRAL ASIA
ANNULAR SOLAR ECLIPSE OF 2010 JANUARY 15

Location Name	Latitude	Longitude	Elev.	First Contact U.T. h m s	P °	V °	Alt °	Second Contact U.T. h m s	P °	V °	Third Contact U.T. h m s	P °	V °	Fourth Contact U.T. h m s	P °	V °	Alt °	Maximum Eclipse U.T. h m s	P °	V °	Alt °	Azm °	Eclip. Mag.	Eclip. Obs.	Umbral Depth	Umbral Durat.
			m																								
KAZAKHSTAN																										
Alma-Ata	43°15'N	076°57'E	775	06:54:49.2	199	200	26	—			—			09:20:06.1	106	80	18	08:10:27.2	152	138	24	198	0.288	0.173		
KYRGYZSTAN																										
Bishkek	42°54'N	074°36'E	—	06:51:24.1	197	201	26	—			—			09:14:46.0	108	84	20	08:05:46.9	152	141	25	194	0.266	0.154		
MONGOLIA																										
Ulaanbaatar	47°55'N	106°53'E	1307	07:26:01.7	218	193	14	—			—			—				08:42:12.0	153	120	5	230	0.547	0.429		
TAJIKISTAN																										
Dusanbe	38°35'N	068°48'E	—	06:33:06.6	198	211	29	—			—			09:05:31.4	108	88	27	07:51:42.8	152	148	30	185	0.259	0.148		
TURKMENISTAN																										
Aschabad	37°57'N	058°23'E	—	06:13:24.0	194	220	24	—			—			08:26:34.2	120	117	31	07:19:44.9	157	169	30	165	0.171	0.081		
UZBEKISTAN																										
Taskent	41°20'N	069°18'E	—	06:41:09.1	195	206	26	—			—			09:02:05.5	111	93	24	07:53:38.0	153	148	27	186	0.229	0.123		

ANNULAR SOLAR ECLIPSE OF 2010 JANUARY 15

TABLE 2.13
LOCAL CIRCUMSTANCES FOR SOUTH ASIA
ANNULAR SOLAR ECLIPSE OF 2010 JANUARY 15

Location Name	Latitude	Longitude	Elev. (m)	First Contact U.T. h m s	P °	V °	Alt °	Second Contact U.T. h m s	P °	V °	Third Contact U.T. h m s	P °	V °	Fourth Contact U.T. h m s	P °	V °	Alt °	Maximum Eclipse U.T. h m s	P °	V °	Alt °	Azm °	Eclip. Mag.	Eclip. Obs.	Umbral Depth	Umbral Durat.
AFGHANISTAN																										
Kabul	34°31'N	069°12'E	1815	06:22:24.6	203	220	32	—			—			09:13:24.3	101	77	29	07:51:13.4	151	147	34	185	0.322	0.203		
BANGLADESH																										
Chittagong	22°20'N	091°50'E	—	06:45:36.1	233	218	45	—			—			10:01:47.9	61	4	17	08:32:53.2	146	104	33	223	0.898	0.825		
Dacca	23°43'N	090°25'E	—	06:44:06.7	230	218	44	—			—			10:00:58.7	64	9	18	08:31:36.9	147	107	33	221	0.844	0.770		
Khulna	22°48'N	089°33'E	—	06:40:38.1	229	218	45	—			—			10:00:10.2	66	9	19	08:29:41.6	146	107	35	220	0.850	0.776		
Rajshahi	24°22'N	088°36'E	—	06:40:46.9	228	219	44	—			—			09:59:24.9	67	14	19	08:29:08.2	147	109	34	218	0.800	0.720		
BHUTAN																										
Thimbu	27°28'N	089°39'E	—	06:47:54.3	225	214	41	—			—			09:59:53.2	71	19	16	08:32:14.3	147	111	31	219	0.751	0.663		
INDIA																										
Ahmadabad	23°02'N	072°37'E	55	05:57:16.2	218	243	42	—			—			09:33:32.6	81	42	35	07:52:21.6	148	137	45	191	0.564	0.450		
Ambasamudram	08°42'N	077°28'E	—	05:37:05.7	238	275	54	07:41:56.6	199	178	07:50:19.7	92	68	09:36:28.8	56	360	41	07:46:08.7	145	123	58	201	0.918	0.843	0.406	08m23s
Aruppukkottai	09°31'N	078°06'E	—	05:40:53.1	238	271	54	07:45:49.7	187	165	07:52:47.4	103	77	09:38:26.1	56	360	40	07:49:18.8	145	123	57	203	0.918	0.842	0.260	06m58s
Bangalore	12°59'N	077°35'E	895	05:46:31.4	233	262	52	—			—			09:38:22.8	62	8	38	07:53:23.1	146	123	53	202	0.846	0.774		
Bombay	18°58'N	072°50'E	8	05:46:43.5	223	254	44	—			—			09:34:30.4	75	33	38	07:48:10.7	147	136	49	191	0.645	0.541		
Calcutta	22°32'N	088°22'E	6	06:37:15.5	229	222	46	—			—			09:36:55.5	64	10	20	08:27:34.8	146	108	36	219	0.835	0.761		
Delhi	28°40'N	077°13'E	—	06:23:15.8	214	225	39	—			—			09:40:59.6	84	44	27	08:09:10.3	148	129	38	200	0.531	0.413		
Hyderabad	17°23'N	078°29'E	—	05:59:24.0	228	248	49	—			—			09:45:17.8	67	16	33	08:00:20.3	146	122	48	204	0.771	0.687		
Jaipur	26°55'N	075°49'E	—	06:15:42.0	215	230	40	—			—			09:39:11.3	83	43	29	08:04:37.6	148	131	40	198	0.542	0.425		
Kanpur	26°28'N	080°21'E	—	06:25:18.6	219	225	42	—			—			09:48:12.2	78	33	25	08:14:56.1	147	122	38	206	0.622	0.514		
Karaikkudi	10°04'N	078°47'E	—	05:44:19.1	237	268	54	07:48:32.1	189	164	07:55:38.4	101	74	09:40:04.7	56	360	39	07:52:05.5	145	119	56	204	0.918	0.842	0.278	07m06s
Kanpur	26°28'N	080°21'E	—	06:25:18.6	219	225	42	—			—			09:48:12.2	78	33	25	08:14:56.1	147	122	38	206	0.622	0.514		
Karaikkudi	10°04'N	078°47'E	—	05:44:19.1	237	268	54	07:48:32.1	189	164	07:55:38.4	101	74	09:40:04.7	56	360	39	07:52:05.5	145	119	56	204	0.918	0.842	0.278	07m06s
Kumbakonam	10°58'N	079°23'E	—	05:48:10.5	236	264	54	07:53:07.5	168	142	07:57:09.8	122	93	09:41:55.6	57	0	37	07:55:08.5	145	118	54	206	0.917	0.842	0.082	04m02s
Lucknow	26°51'N	080°55'E	122	06:27:28.0	219	224	42	—			—			09:48:58.7	78	33	25	08:16:22.5	147	121	37	207	0.623	0.516		
Madras	13°05'N	080°17'E	16	05:55:32.4	234	256	53	—			—			09:45:13.9	59	3	35	08:00:40.5	145	116	51	208	0.892	0.823		
Madurai	09°56'N	078°07'E	—	05:41:47.0	237	270	54	07:48:24.5	164	140	07:51:42.9	127	102	09:38:56.7	57	1	40	07:50:03.5	145	121	56	203	0.918	0.842	0.053	03m18s
Nagappattinam	10°46'N	079°50'E	—	05:49:17.7	237	263	55	07:51:59.2	197	170	07:59:58.9	93	63	09:42:19.8	56	359	37	07:55:59.5	145	116	54	207	0.917	0.842	0.384	08m00s
Nagercoil	08°10'N	077°26'E	—	05:35:56.2	239	276	54	07:40:10.6	216	196	07:50:02.0	75	50	09:44:55.4	55	358	42	07:45:07.1	145	123	59	201	0.918	0.843	0.669	09m51s
Nagpur	21°09'N	079°06'E	—	06:09:58.8	224	238	46	—			—			09:47:09.6	71	23	30	08:07:46.4	146	122	44	205	0.706	0.611		
New Delhi	28°36'N	077°12'E	212	06:23:04.2	214	225	39	—			—			09:41:00.7	84	44	27	08:09:06.0	148	129	38	200	0.532	0.414		
Palayankottai	08°43'N	077°44'E	—	06:38:02.4	239	274	54	07:42:22.2	206	185	07:51:27.6	84	59	09:36:54.5	56	359	41	07:46:55.5	145	122	58	202	0.918	0.843	0.514	09m05s
Patna	25°36'N	085°07'E	—	06:34:42.2	223	221	43	—			—			09:55:27.8	72	22	22	08:23:15.1	147	114	36	213	0.717	0.623		
Pudukkottai	10°23'N	078°49'E	—	05:45:04.3	237	267	54	07:50:10.4	175	150	07:55:13.8	115	88	09:40:29.0	57	0	38	07:52:42.1	145	119	55	205	0.918	0.842	0.130	05m03s
Pune	18°32'N	073°52'E	—	05:48:34.2	224	253	45	—			—			09:40:50.4	73	29	37	07:50:44.5	147	134	49	193	0.670	0.570		
Quilon	08°53'N	076°36'E	—	05:34:31.1	238	276	53	—			—			09:35:19.6	57	2	42	07:43:58.5	146	126	58	198	0.915	0.842		
Rajpalaiyam	09°27'N	077°34'E	—	05:38:56.2	237	272	54	07:45:29.9	165	143	07:49:29.9	125	102	09:37:32.6	57	1	41	07:47:42.9	145	123	57	201	0.918	0.842	0.060	03m34s
Sivakasi	09°27'N	077°49'E	—	05:39:47.2	238	272	54	07:45:31.2	179	157	07:51:18.4	111	87	09:37:32.6	57	0	40	07:48:27.2	145	122	57	202	0.918	0.842	0.170	05m47s
Thanjavur	10°48'N	079°09'E	—	05:47:02.9	236	265	54	07:52:28.2	166	139	07:56:03.7	124	97	09:41:25.1	57	0	38	07:54:15.8	145	118	55	205	0.918	0.842	0.064	03m36s
Tirunelveli	08°44'N	077°42'E	—	05:37:55.5	238	274	54	07:42:22.4	205	184	07:51:19.6	86	61	09:36:11.6	56	359	41	07:46:51.6	145	122	58	202	0.918	0.843	0.491	08m57s
Trivandrum	08°29'N	076°55'E	—	05:34:47.5	238	276	54	07:40:33.5	189	170	07:47:49.1	102	79	09:35:20.7	56	0	42	07:44:11.6	146	125	59	199	0.918	0.843	0.279	07m16s
Tuticorin	09°47'N	078°08'E	—	05:39:32.6	239	273	55	07:43:16.8	215	193	07:53:00.3	75	49	09:37:35.4	55	358	41	07:48:09.3	145	121	58	203	0.918	0.843	0.655	09m44s
Virudunagar	09°36'N	077°58'E	—	05:40:35.9	237	271	54	07:46:17.2	178	155	07:51:32.9	112	88	09:38:20.0	57	0	40	07:49:05.1	145	122	57	202	0.918	0.842	0.158	05m36s
Vishakhapatnam	17°42'N	083°18'E	—	06:14:29.9	231	239	51	—			—			09:52:15.1	62	7	28	08:13:35.9	145	112	44	213	0.850	0.777		
MALDIVES																										
Male	04°10'N	073°30'E	—	05:15:21.5	244	296	51	07:20:20.8	245	242	07:31:06.0	49	40	09:23:15.1	54	359	50	07:25:44.2	327	321	65	186	0.919	0.844	0.857	10m45s
NEPAL																										
Kathmandu	27°43'N	085°19'E	1348	06:39:03.6	221	218	41	—			—			09:55:01.9	75	27	20	08:25:19.0	147	116	34	213	0.677	0.577		
PAKISTAN																										
Faisalabad	31°25'N	073°05'E	—	06:21:25.2	208	223	36	—			—			09:28:42.7	92	60	29	08:00:19.0	150	138	36	193	0.422	0.299		
Islamabad	33°42'N	073°10'E	—	06:27:25.4	206	218	34	—			—			09:26:10.3	96	65	27	08:01:33.6	150	139	34	178	0.386	0.263		
Karachi	24°52'N	067°03'E	4	05:48:11.0	213	246	36	—			—			09:16:43.8	90	62	39	07:36:19.5	151	152	44	173	0.452	0.330		
Lahore	31°35'N	074°18'E	—	06:24:20.7	209	221	36	—			—			09:31:38.9	91	58	28	08:03:32.5	149	136	36	195	0.436	0.314		
SRI LANKA																										
Colombo	06°56'N	079°51'E	7	05:42:06.1	242	275	57	—			07:59:42.2	53	22	09:37:24.6	51	351	40	07:49:34.8	325	296	58	207	0.902	0.833		
Jaffna	09°40'N	080°00'E	—	05:47:41.8	238	266	56	07:49:32.7	236	209	07:59:42.2	53	22	09:41:17.0	54	356	38	07:54:38.4	325	296	55	207	0.918	0.842	0.975	10m09s

ANNULAR SOLAR ECLIPSE OF 2010 JANUARY 15

TABLE 2.14
LOCAL CIRCUMSTANCES FOR SOUTHEAST ASIA
ANNULAR SOLAR ECLIPSE OF 2010 JANUARY 15

Location Name	Latitude	Longitude	Elev.	First Contact U.T. h m s	P °	V °	Alt °	Second Contact U.T. h m s	P °	V °	Third Contact U.T. h m s	P °	V °	Fourth Contact U.T. h m s	P °	V °	Alt °	Maximum Eclipse U.T. h m s	P °	V °	Alt °	Azm °	Eclip. Mag.	Eclip. Obs.	Umbral Depth	Umbral Durat.
BRUNEI DARUSSALAM			m																							
Bandar Seri Beg*	04°56'N	114°55'E	3	07:45:20.8	286	219	35	—			—			09:33:42.6	11	291	11	08:42:35.2	329	254	23	245	0.235	0.129		
BURMA (MYANMAR)																										
Chauk	20°54'N	094°50'E	—	06:51:16.7	237	217	45	08:32:31.5	269	223	08:39:46.7	24	336	10:02:59.7	57	357	15	08:36:09.4	327	280	32	227	0.913	0.834	0.460	07m15s
Lashio	22°56'N	097°45'E	—	07:00:15.4	238	212	42	08:37:31.4	272	224	08:44:26.2	23	334	10:04:56.2	57	357	12	08:40:59.0	327	279	27	229	0.912	0.832	0.434	06m55s
Magway	20°09'N	094°55'E	—	06:50:39.4	238	218	46	08:34:37.1	311	264	08:36:53.6	342	294	10:02:40.2	55	355	16	08:35:44.9	326	279	32	227	0.913	0.834	0.035	02m16s
Mandalay	22°00'N	096°05'E	77	06:55:30.1	237	215	44	08:34:38.4	263	217	08:42:17.2	31	343	10:03:59.3	57	357	14	08:38:28.2	327	280	30	227	0.913	0.833	0.560	07m39s
Maymyo	22°02'N	096°28'E	—	06:56:26.6	238	214	43	08:35:27.7	271	225	08:42:27.6	23	334	10:04:09.2	57	357	14	08:38:57.9	327	279	29	228	0.913	0.833	0.434	07m00s
Meiktila	20°52'N	095°52'E	—	06:53:48.7	239	216	45	08:36:11.6	309	261	08:38:48.2	345	296	10:03:25.3	56	355	14	08:37:29.5	327	279	31	228	0.913	0.834	0.048	02m37s
Monywa	22°05'N	095°08'E	—	06:53:19.2	236	215	44	08:33:00.2	239	194	08:41:34.8	54	8	10:03:36.0	58	359	14	08:37:18.1	327	281	30	226	0.913	0.834	0.961	08m35s
Myingyan	21°28'N	095°23'E	—	06:53:15.4	237	216	45	08:33:25.6	264	218	08:41:03.7	30	342	10:03:29.0	57	357	15	08:37:15.0	327	280	31	227	0.913	0.834	0.545	07m38s
Myitkyina	25°23'N	097°24'E	—	07:01:43.3	235	211	39	08:39:18.9	181	136	08:43:57.9	114	68	10:05:17.4	61	4	11	08:41:38.2	148	102	26	228	0.912	0.832	0.168	04m39s
Pakokku	21°20'N	095°05'E	—	06:52:22.6	237	216	45	08:32:50.8	261	215	08:40:41.1	32	345	10:03:17.7	57	357	15	08:36:46.4	327	280	31	227	0.913	0.834	0.590	07m50s
Shwebo	22°34'N	095°42'E	—	06:55:10.9	236	214	43	08:34:01.8	237	191	08:42:34.0	57	10	10:04:00.5	58	359	14	08:38:18.5	327	280	29	227	0.913	0.833	0.998	08m32s
Sittwe	20°09'N	092°54'E	—	06:45:26.8	236	220	47	08:28:35.5	246	202	08:37:13.7	46	360	10:01:40.3	57	358	18	08:32:55.2	326	281	34	225	0.914	0.835	0.823	08m38s
Yangon	16°47'N	096°10'E	—	06:50:20.9	244	219	49	—			—			10:00:46.6	49	345	17	08:34:44.5	326	275	34	229	0.817	0.740		
Yenangyaung	20°28'N	094°52'E	—	06:50:52.8	238	217	45	08:33:12.3	288	241	08:38:36.4	5	318	10:02:48.5	56	355	16	08:35:54.3	326	279	32	227	0.913	0.834	0.221	05m24s
CAMBODIA																										
Phnum Pénh	11°33'N	104°55'E	12	07:12:28.0	261	215	46	—			—			09:54:22.9	33	321	13	08:40:19.8	327	264	29	238	0.548	0.432		
LAOS																										
Vientiane	17°58'N	102°36'E	170	07:08:17.0	249	213	43	—			—			10:02:44.3	46	340	10	08:43:18.7	327	271	27	234	0.738	0.647		
INDONESIA																										
Bandung	06°54'S	107°36'E	—	07:39:09.0	301	226	49	—			—			08:54:53.9	351	266	31	08:18:21.8	326	245	40	248	0.084	0.028		
Bandung	06°54'S	107°36'E	—	07:39:09.0	301	226	49	—			—			08:54:53.9	351	266	31	08:18:21.8	326	245	40	248	0.084	0.028		
Brebes	06°53'S	109°03'E	—	07:47:44.1	306	228	48	—			—			08:20:17.5	348	262	31	08:20:17.5	326	245	38	248	0.060	0.017		
Cianjur	06°49'S	107°08'E	—	07:36:14.4	300	225	50	—			—			08:56:27.3	353	268	31	08:17:52.3	326	245	40	248	0.093	0.033		
Cibinong	06°27'S	106°51'E	—	07:33:39.0	298	223	51	—			—			08:59:02.7	354	270	31	08:18:06.2	326	246	40	248	0.106	0.040		
Cilacap	07°44'S	109°00'E	—	07:51:24.6	309	230	45	—			—			08:45:04.9	344	259	33	08:18:48.7	326	244	39	249	0.042	0.010		
Ciledug	06°14'S	108°44'E	—	07:45:51.1	305	227	46	—			—			08:51:54.9	348	263	31	08:19:51.4	326	245	38	248	0.085	0.019		
Cimahi	06°53'S	107°32'E	—	07:38:42.8	301	226	49	—			—			08:55:09.3	351	266	31	08:18:18.1	326	245	40	248	0.085	0.029		
Cipazay	07°03'S	107°43'E	—	07:40:19.7	302	226	48	—			—			08:53:43.1	350	265	32	08:18:16.0	326	245	40	248	0.078	0.025		
Depok	06°24'S	106°50'E	—	07:33:25.6	298	224	51	—			—			08:59:20.4	354	270	31	08:18:09.9	326	246	40	248	0.107	0.041		
Garut	07°13'S	107°54'E	—	07:42:00.5	303	227	48	—			—			08:52:13.0	349	264	32	08:18:13.9	326	245	40	248	0.072	0.022		
Indramayu	06°20'S	108°19'E	—	07:41:22.9	302	226	47	—			—			08:56:28.9	351	266	30	08:20:15.3	326	245	39	248	0.084	0.028		
Jakarta	06°10'S	106°48'E	8	07:32:38.9	297	224	51	—			—			09:00:36.9	355	271	31	08:18:30.7	326	246	40	247	0.113	0.044		
Majalaya	07°03'S	107°45'E	—	07:40:31.4	302	226	48	—			—			08:53:38.1	350	265	32	08:18:18.7	326	245	40	248	0.078	0.025		
Medan	03°35'N	098°40'E	—	06:49:48.9	265	226	59	—			—			09:40:03.6	26	311	25	08:22:21.3	325	261	42	237	0.473	0.352		
Pringsewu	05°24'S	104°55'E	—	07:21:26.8	291	223	54	—			—			09:07:18.4	1	277	30	08:17:10.1	326	248	42	246	0.162	0.075		
Purwokerto	07°25'S	109°14'E	—	07:51:23.5	308	229	45	—			—			08:46:39.7	345	259	32	08:19:38.5	326	244	38	249	0.045	0.011		
Semarang	06°58'S	110°25'E	—	07:57:16.4	311	231	42	—			—			08:45:36.1	343	258	31	08:21:51.7	327	244	36	249	0.035	0.008		
Serang	06°07'S	106°09'E	—	07:29:06.9	296	224	52	—			—			09:02:03.3	357	272	31	08:17:41.9	326	247	41	247	0.125	0.051		
Tangerang	06°11'S	106°37'E	—	07:31:43.2	297	224	51	—			—			09:00:52.6	355	271	31	08:18:13.9	326	246	40	247	0.116	0.045		
Tasikmalaya	07°20'S	108°12'E	—	07:44:11.3	305	227	47	—			—			08:50:37.5	348	263	32	08:18:26.1	326	244	39	248	0.064	0.019		
Tembilahan	00°19'S	103°09'E	—	07:08:07.9	277	221	56	—			—			09:27:40.4	14	294	25	08:22:56.7	325	254	40	242	0.305	0.188		
MACAU																										
Macau	22°14'N	113°35'E	—	07:31:47.1	255	208	29	—			—			—				08:53:51.7	330	270	14	240	0.680	0.580		
MALAYSIA																										
Kuala Lumpur	03°10'N	101°42'E	34	07:01:22.2	269	222	56	—			—			09:38:09.1	22	305	23	08:26:10.1	325	258	39	239	0.409	0.286		
PHILIPPINES																										
Cebu	10°18'N	123°54'E	—	07:59:55.6	288	219	22	—			—			09:39:46.3	14	295	0	08:52:18.7	331	256	10	246	0.246	0.137		
Davao	07°04'N	125°36'E	27	08:09:57.6	291	224	20	—			—			09:28:05.0	4	282	2	08:50:28.2	331	253	11	247	0.144	0.063		
Manila	14°35'N	121°00'E	15	07:49:03.6	276	214	25	—			—			—				08:53:56.5	331	261	11	245	0.390	0.267		
Quezon City	14°38'N	121°03'E	—	07:49:06.3	275	214	24	—			—			—				08:53:59.3	331	261	11	245	0.391	0.268		
SINGAPORE																										
Singapore	01°17'N	103°51'E	10	07:10:08.3	276	220	54	—			—			09:32:00.4	16	297	23	08:26:20.2	326	255	38	242	0.329	0.210		
TAIWAN																										
T'aipei	25°03'N	121°30'E	6	07:42:28.6	259	207	20	—			—			—				08:57:28.7	332	271	5	244	0.653	0.549		
THAILAND																										
Bangkok	13°45'N	100°31'E	16	07:00:30.9	253	217	48	—			—			09:58:08.3	41	332	15	08:37:30.0	326	269	32	234	0.673	0.573		
VIETNAM																										
Ha Noi	21°02'N	105°51'E	6	07:16:54.0	249	209	37	—			—			10:05:03.0	48	343	6	08:48:07.1	328	273	21	236	0.761	0.673		
Ho Chi Minh	10°45'N	106°40'E	10	07:17:28.5	264	215	44	—			—			09:52:29.6	31	317	12	08:41:20.1	327	262	28	239	0.501	0.381		

ANNULAR SOLAR ECLIPSE OF 2010 JANUARY 15

TABLE 2.15
LOCAL CIRCUMSTANCES FOR CHINA, NORTH AND SOUTH KOREA
ANNULAR SOLAR ECLIPSE OF 2010 JANUARY 15

Location Name	Latitude	Longitude	Elev.	First Contact U.T. h m s	P °	V °	Alt °	Second Contact U.T. h m s	P °	V °	Third Contact U.T. h m s	P °	V °	Fourth Contact U.T. h m s	P °	V °	Alt °	Maximum Eclipse U.T. h m s	P °	V °	Alt °	Azm °	Eclip. Mag.	Eclip. Obs.	Umbral Depth	Umbral Durat.
CHINA			m																							
Baoshan	25°09'N	099°09'E	—	07:05:08.7	237	210	39	08:39:20.5	232	186	08:47:35.0	63	15	10:05:54.6	60	1	9	08:43:28.3	148	100	25	229	0.912	0.831	0.905	08m14s
Beijing	39°55'N	116°25'E	—	07:32:48.0	234	198	15	—	—	—	—	—	—	—	—	—	—	08:52:28.4	153	108	3	239	0.823	0.743	—	—
Changchun	43°53'N	125°19'E	—	07:37:11.8	235	197	7	—	—	—	—	—	—	—	—	—	—	08:24 Set	—	—	0	241	0.568	0.453	—	—
Chengdu	30°39'N	104°04'E	—	07:17:24.1	235	204	30	—	—	—	—	—	—	—	—	—	—	08:48:50.3	150	103	17	232	0.895	0.819	—	—
Chongqing	29°34'N	106°35'E	261	07:20:53.1	239	204	29	08:46:47.9	239	191	08:54:37.4	60	10	10:06:53.2	65	9	3	08:50:43.1	150	101	15	234	0.910	0.827	0.994	07m50s
Chuxiong	25°02'N	101°30'E	—	07:09:48.9	239	209	37	08:43:13.7	289	240	08:46:22.6	8	318	10:07:28.2	61	9	3	08:45:48.0	328	279	23	231	0.911	0.830	0.226	05m09s
Dongchuan	26°10'N	103°01'E	—	07:13:21.6	239	207	35	08:44:33.0	281	232	08:50:27.5	16	326	10:06:30.6	58	358	7	08:47:30.3	329	279	21	232	0.911	0.829	0.323	08m06s
Dukou	26°40'N	101°39'E	—	07:11:07.4	237	207	35	08:42:19.4	236	189	08:50:25.8	61	12	10:06:58.7	58	359	6	08:46:23.1	148	100	21	231	0.911	0.830	0.963	08m06s
Enshi	30°17'N	109°19'E	—	07:25:05.8	240	203	27	08:48:57.3	265	215	08:55:57.4	36	345	10:06:45.2	60	2	6	08:52:27.6	330	280	13	236	0.909	0.826	0.588	07m00s
Fuling	29°42'N	107°21'E	—	07:22:05.4	239	204	29	08:47:23.0	249	200	08:55:04.7	51	1	10:07:30.5	61	3	1	08:51:14.2	330	280	15	235	0.909	0.827	0.848	07m42s
Guanghua	32°25'N	111°36'E	—	07:28:21.0	240	202	23	08:49:39.7	236	187	08:57:12.7	66	16	—	—	—	—	08:53:26.6	151	101	10	237	0.908	0.825	0.910	07m33s
Guangzhou	23°06'N	113°16'E	18	07:31:06.3	254	207	29	—	—	—	—	—	—	—	—	—	—	08:53:56.2	330	271	13	240	0.706	0.610	—	—
Harbin	45°45'N	126°41'E	145	07:36:57.9	234	197	5	—	—	—	—	—	—	—	—	—	—	08:13 Set	—	—	0	240	0.432	0.308	—	—
Heze	35°17'N	115°27'E	—	07:32:32.9	239	201	19	08:51:23.5	204	155	08:57:14.3	100	51	—	—	—	—	08:54:19.0	152	103	5	239	0.907	0.823	0.384	05m51s
Jinan	36°40'N	116°57'E	—	07:33:46.4	239	200	17	08:53:35.7	163	115	08:54:58.8	142	93	—	—	—	—	08:54:16.9	153	104	4	240	0.907	0.822	0.018	01m23s
Jingmen	31°00'N	112°09'E	—	07:29:00.0	242	203	24	08:51:44.6	295	245	08:56:07.8	71	315	—	—	—	—	08:53:56.0	331	280	10	238	0.908	0.825	0.186	04m23s
Junxian	32°31'N	111°30'E	—	07:28:13.9	239	202	23	08:49:37.8	231	182	08:57:06.4	71	21	—	—	—	—	08:53:22.4	151	102	10	237	0.908	0.825	0.830	07m29s
Kaifeng	34°51'N	114°21'E	—	07:31:27.7	239	201	21	08:51:14.6	200	153	08:57:08.3	82	32	—	—	—	—	08:54:02.8	152	103	5	240	0.907	0.824	0.648	05m36s
Laiwu	36°12'N	117°40'E	—	07:34:32.4	240	200	16	08:51:40.0	200	159	08:55:44.2	97	47	—	—	—	—	08:54:42.2	153	103	3	241	0.907	0.823	0.339	05m04s
Leshan	29°34'N	103°45'E	—	07:16:23.3	236	205	31	08:46:38.3	178	131	08:50:32.0	120	72	—	—	—	—	08:48:34.9	149	102	18	232	0.910	0.828	0.127	03m54s
Lianyungang	34°39'N	119°16'E	—	07:36:20.7	243	201	16	08:52:52.8	281	230	08:58:32.8	24	332	—	—	—	—	08:55:42.8	333	281	3	242	0.906	0.822	0.374	05m40s
Lüda	38°53'N	121°35'E	—	07:36:55.5	239	199	12	—	—	—	—	—	—	—	—	—	—	08:54 Set	—	—	0	243	0.900	0.818	—	—
Luzhou	28°54'N	105°27'E	—	07:18:51.6	238	205	31	08:45:55.3	240	191	08:53:48.5	59	10	10:07:24.3	61	3	2	08:49:52.4	329	280	17	233	0.910	0.828	0.998	07m53s
Nanjing	32°03'N	118°47'E	—	07:34:36.0	246	202	18	—	—	—	—	—	—	—	—	—	—	08:56:18.6	332	279	4	242	0.864	0.787	—	—
Nanyang	33°00'N	112°32'E	—	07:29:29.8	240	202	22	08:50:02.4	233	183	08:57:29.7	70	20	—	—	—	—	08:53:46.4	151	102	9	238	0.908	0.824	0.849	07m27s
Neijiang	29°35'N	105°03'E	—	07:18:29.1	237	204	31	08:46:07.4	211	163	08:53:33.0	88	39	—	—	—	—	08:49:35.8	150	101	17	233	0.910	0.828	0.521	06m56s
Pingdingshan	33°45'N	113°17'E	—	07:30:15.2	239	201	21	08:50:24.6	217	172	08:57:26.3	82	32	—	—	—	—	08:53:55.7	152	102	7	239	0.908	0.824	0.468	07m02s
Puyang	35°42'N	114°59'E	—	07:32:02.3	238	200	19	08:52:49.6	170	122	08:59:00.9	134	85	—	—	—	—	08:55:59.6	152	104	6	239	0.907	0.823	0.051	02m21s
Qingdao	36°05'N	120°19'E	—	07:36:05.5	242	201	15	08:51:48.4	248	198	08:58:00.9	58	7	—	—	—	—	08:55:25.0	333	283	2	242	0.906	0.821	0.913	07m13s
Rizhao	35°27'N	119°29'E	—	07:36:19.0	242	201	16	08:51:57.3	257	207	08:58:59.7	49	358	—	—	—	—	08:55:28.8	333	282	2	242	0.906	0.821	0.754	07m02s
Shanghai	31°14'N	121°28'E	5	07:39:36.2	250	203	16	—	—	—	—	—	—	—	—	—	—	08:57:08.3	333	277	3	243	0.813	0.732	—	—
Shenyang	41°48'N	123°27'E	42	07:37:04.2	237	198	9	—	—	—	—	—	—	—	—	—	—	08:38 Set	—	—	0	242	0.731	0.638	—	—
Shuicheng	26°41'N	104°50'E	—	07:16:56.4	240	206	33	08:48:30.3	320	269	08:49:46.8	338	288	—	—	—	—	08:49:08.1	329	279	19	234	0.910	0.827	0.013	01m17s
Suining	30°31'N	105°34'E	—	07:19:38.9	236	204	29	08:47:39.8	186	138	08:52:17.5	114	66	—	—	—	—	08:49:58.5	150	102	16	233	0.910	0.829	0.191	04m38s
Tai'an	36°12'N	117°07'E	—	07:34:00.5	239	200	17	08:51:53.6	198	149	08:57:09.7	107	58	—	—	—	—	08:54:31.6	152	104	4	240	0.907	0.822	0.302	05m16s
Taiyuan	37°55'N	112°30'E	—	07:29:30.8	233	199	19	—	—	—	—	—	—	—	—	—	—	08:52:05.3	152	107	6	237	0.828	0.749	—	—
Tianjin	39°08'N	104°38'E	4	07:33:33.1	236	199	15	—	—	—	—	—	—	—	—	—	—	08:53:09.3	153	107	3	240	0.850	0.773	—	—
Wanxian	30°52'N	108°22'E	—	07:23:11.4	239	203	27	08:48:01.4	231	182	08:55:38.8	70	20	10:07:18.2	61	4	3	08:51:50.5	150	101	13	235	0.909	0.826	0.831	07m37s
Weifang	36°42'N	119°04'E	—	07:35:37.2	241	200	15	08:51:43.9	212	163	08:57:59.7	94	44	—	—	—	—	08:54:51.9	153	104	2	242	0.906	0.821	0.489	06m16s
Wuhan	30°36'N	114°17'E	23	07:31:44.1	244	201	23	—	—	—	—	—	—	—	—	—	—	08:54:57.8	331	279	9	239	0.880	0.805	—	—
Xiaguan	25°34'N	100°14'E	—	07:07:39.1	237	209	37	08:40:37.3	242	194	08:48:48.7	55	6	10:06:17.4	59	1	8	08:44:43.5	328	280	23	230	0.911	0.831	0.938	08m11s
Xi'an	34°15'N	108°52'E	—	07:25:11.8	235	201	24	—	—	—	—	—	—	—	—	—	—	08:51:33.7	151	104	11	235	0.872	0.797	—	—
Xiangfan	32°03'N	115°01'E	—	07:28:51.3	240	200	23	08:50:01.9	254	204	08:57:24.0	49	358	—	—	—	—	08:53:42.9	331	281	10	238	0.908	0.825	0.784	07m23s
Xintai	35°54'N	117°44'E	—	07:34:38.0	240	200	17	08:51:27.7	219	170	08:58:11.9	86	36	—	—	—	—	08:54:50.0	153	103	4	241	0.907	0.822	0.604	06m44s
Xuzhou	34°16'N	117°11'E	—	07:34:26.4	242	201	18	08:51:47.3	262	212	08:58:41.5	42	351	—	—	—	—	08:55:14.6	332	282	5	241	0.907	0.822	0.658	06m54s
Yibin	28°47'N	104°38'E	—	07:17:28.2	238	205	31	08:45:20.3	228	180	08:53:07.5	70	21	10:07:18.2	61	4	3	08:49:14.4	149	101	18	233	0.910	0.828	0.811	07m47s
Yichang	30°42'N	111°17'E	—	07:27:51.1	241	203	25	08:50:58.3	288	238	08:56:05.0	13	322	—	—	—	—	08:53:31.6	331	280	11	237	0.908	0.825	0.261	05m07s
Zaozhuang	34°53'N	117°34'E	—	07:34:41.4	241	201	17	08:51:30.4	248	198	08:58:10.4	56	6	—	—	—	—	08:55:09.5	332	282	4	241	0.907	0.822	0.996	07m18s
Zhaotong	27°19'N	103°48'E	—	07:15:23.6	239	206	33	08:44:36.9	258	209	08:52:10.5	40	350	10:07:12.1	60	1	4	08:48:24.0	329	280	19	233	0.910	0.829	0.678	07m34s
Zhengzhou	34°48'N	113°39'E	—	07:30:43.4	238	201	20	08:51:35.0	187	139	08:55:58.5	116	67	—	—	—	—	08:53:46.6	152	103	7	238	0.907	0.824	0.189	04m23s
Zibo	36°47'N	118°01'E	—	07:34:41.9	239	200	16	08:52:19.6	190	141	08:56:46.4	115	66	—	—	—	—	08:54:32.8	153	104	3	241	0.906	0.822	0.205	04m27s
Zigong	29°24'N	104°47'E	—	07:17:59.0	237	205	31	08:45:52.7	212	164	08:52:53.5	87	38	10:07:15.8	62	5	3	08:49:23.3	149	101	17	233	0.910	0.828	0.535	07m01s
HONG KONG																										
New Kowloon	22°20'N	114°10'E	—	07:32:45.6	256	208	28	—	—	—	—	—	—	—	—	—	—	08:54:11.8	330	270	13	240	0.675	0.574	—	—
KOREA, NORTH																										
P'yongyang	39°01'N	125°45'E	29	07:39:35.5	242	200	9	08:36:31.0	—	—	—	—	—	—	—	—	—	08:37 Set	—	—	0	243	0.710	0.613	—	—
KOREA, SOUTH																										
Inch'on	37°28'N	126°38'E	—	07:40:52.9	245	201	9	08:36:58.7	—	—	—	—	—	—	—	—	—	08:37 Set	—	—	0	243	0.701	0.603	—	—
Pusan	35°06'N	129°03'E	2	07:43:42.1	250	202	8	—	—	—	—	—	—	—	—	—	—	08:33 Set	—	—	0	244	0.613	0.502	—	—
Seoul	37°33'N	126°58'E	10	07:41:02.7	245	201	9	08:35:26.1	—	—	—	—	—	—	—	—	—	08:35 Set	—	—	0	243	0.681	0.580	—	—

Table 2.16: Climate Statistics for January along the Annular Eclipse Path

January Climate Statistics	Percent of possible sunshine	Percent Frequency of Sky Condition						Calculated Cloudiness	Prevailing wind	January Rainfall	Days with Rain	Average January High	Average January Low
		Clear	Few	Scattered	Broken	Overcast	Obscured	%		mm		°C	°C
Central African Republic													
Bouar *		32.5	14.4	18.7	30.6	2.9	1	38					
Bossangoa *		29.9	21.4	17.1	27.8	3.9	0	36	calm				
Bangui *	59	20.4	10.5	21.9	40.8	4.7	1.7	49	calm	25	3	32	20
Mobaye *		16	15	15.5	46	4.5	3	53	calm				
Democratic Republic of the Congo													
Mbandaka		11.6	10.9	5.8	51.4	18.1	2.2	65					
Buta		15.6	11.1	6.7	40	26.7	0	63					
Kisangani	50	2.6	5.2	9.1	58.4	23.4	1.3	75		53	6	31	21
Uganda													
Soroti *		2	9.8	20.6	58.8	8.3	0.5	65					
Arua *	63	2.6	9.3	24.4	63.2	0.5	0	62		12	6	30	
Masindi*		5.7	21.1	22.8	45.5	4.9	0	54			5	31	17
Entebbe Airport *	62	1.2	8.9	14.3	52.9	22.8	0	72	S	100	9	29	14
Kampala *	41	0	4.5	8.9	65.2	21.4	0	77		46	9	28	18
Jinja *		3.8	13	22.1	43.5	16.8	0.8	63					
Kenya													
Kusumu	74	3.6	24.9	22.3	45.2	3.9	0.2	53	E	48	6	29	18
Kitale*	69	7.9	30.8	17.2	40.5	3.1	0.5	48	E - NE	20	8	27	11
Eldoret*		6.9	28	20.5	42.4	2.1	0	49	E - NE	34	3	27	5
Lodwar	85	15.8	32.8	20.5	29.8	1.1	0	38	calm	9	1	36	22
Nairobi*	76	7.5	16.3	20	50.1	6.2	0	57	N - NE	58	4	26	11
Garissa*	67	0.9	2.4	8.7	71.9	16	0	76	S - SE			37	23
Mombasa	71	0	5	26	63.9	5.2	0	67	calm	34	3	33	22
Lamu* (coast)		0	4.3	25.2	69	1.5	0	67	E		1	30	26
Seychelles													
Seychelles Airport (Victoria)	40	0	2	15.6	68.9	13.4	0	74	W	379	17	30	24
Maldives													
Hanimaadhoo		0.5	51.4	15.6	30.9	1.7	0	41					
Male*	67	0	24.7	24.9	44.9	5.5	0	55		75	3	30	25
Kadhdhoo		0	28.2	19.5	43.1	9.2	0	56					
India													
Minicoy	77	1.7	23.6	36.7	34.5	3.5	0	50	NE	13	1.7	31	23
Kollam (Quilon)										20	1.2		
Thiruvananthapuram	72	4.1	29.7	28.8	35.3	2	0	47	SW	23	1.6	32	22
Madurai*	71	11.4	36.5	18	26.9	7.2	0	42		16	1	30	21
Cuddalore		7	27.7	31.3	30.4	3.6	0	45				29	21
Sri Lanka													
Colombo	68	3.2	20.5	25.6	43.6	7.1	0	56	NE	58	5	31	22
Jaffna										70	4	29	23
Trincomalee		0	12.5	16.3	59.6	11.5	0	67	NE	116	7	28	24
Bangladesh													
Cox's Bazar		69.5	14.5	9.9	5.3	0.8	0	12		9		26	15
Burma													
Sittwe*	79	56.5	17.7	13.8	10.8	1.3	0	19		11	<1	28	15
Mindat*		32.2	32.7	15.9	15.9	1.9	1.4	28		4		19	9
Kalewa*		11.1	48.6	24.1	11.1	4.6	0.5	32		2		26	13
Mandalay*	82	51.3	32.6	8.5	6.2	1.4	0	15		4	0.1	29	13
Lashio*	75	19.9	42.1	19.2	14.8	2.6	1.5	31		8	1	25	5
China													
Tengchong*	74	11.8	46.8	8.7	28.3	4.4	0	38		16	2.8	16	1
Xichang*	72	25.5	47.1	3.9	16.2	7.4	0	29		6	1.2	16	4
Kunming	69	24.1	32.5	8.4	25.3	9.7	0	38	SW	12	2.2	15	2
Chongqing*		15.8	2	0.7	9.9	71.6	0	80		20	10	10	6
Yichang*	29	21.2	5.7	2.3	21.8	48.8	0.2	68		19	3.8	9	2
Zengzhou*	51	33.4	9.6	3.4	24.6	27	2	51		9	1.8	6	-5
Qingdao*	60	36.8	12.3	5.4	19.2	23.9	2.4	46		11	1.9	3	-4

Table 2.15: Climate statistics for January are given for sites along the annular eclipse path. Location names with a star (*) are in the annular eclipse path.

Explanation of columns:

Percent of possible sunshine: the average number of daily sunshine hours recorded at the station for the month divided by the duration of daylight. This is the best estimate of the probability of seeing the eclipse.

Sky condition: Clear means no cloud, few means 1 or 2 oktas (eights) of sky cover, scattered means 3 or 4 oktas, broken means 5 to 7 oktas, and overcast means no sky visible at all. Obscured refers to a fog layer through which the sky cannot be seen; it is treated as overcast.

Calculated cloudiness: an average cloudiness derived from the frequency and sky cover in the sky condition columns.

Prevailing wind: the most common direction from which the wind blows during the month.

Rainfall: average monthly rainfall

Days with rain: average number of days in January with 0.2 mm of rain or more.

Average high, low: average daily maximum and minimum temperatures for January.

Figure 2.1: Orthographic Projection Map of The Eclipse Path

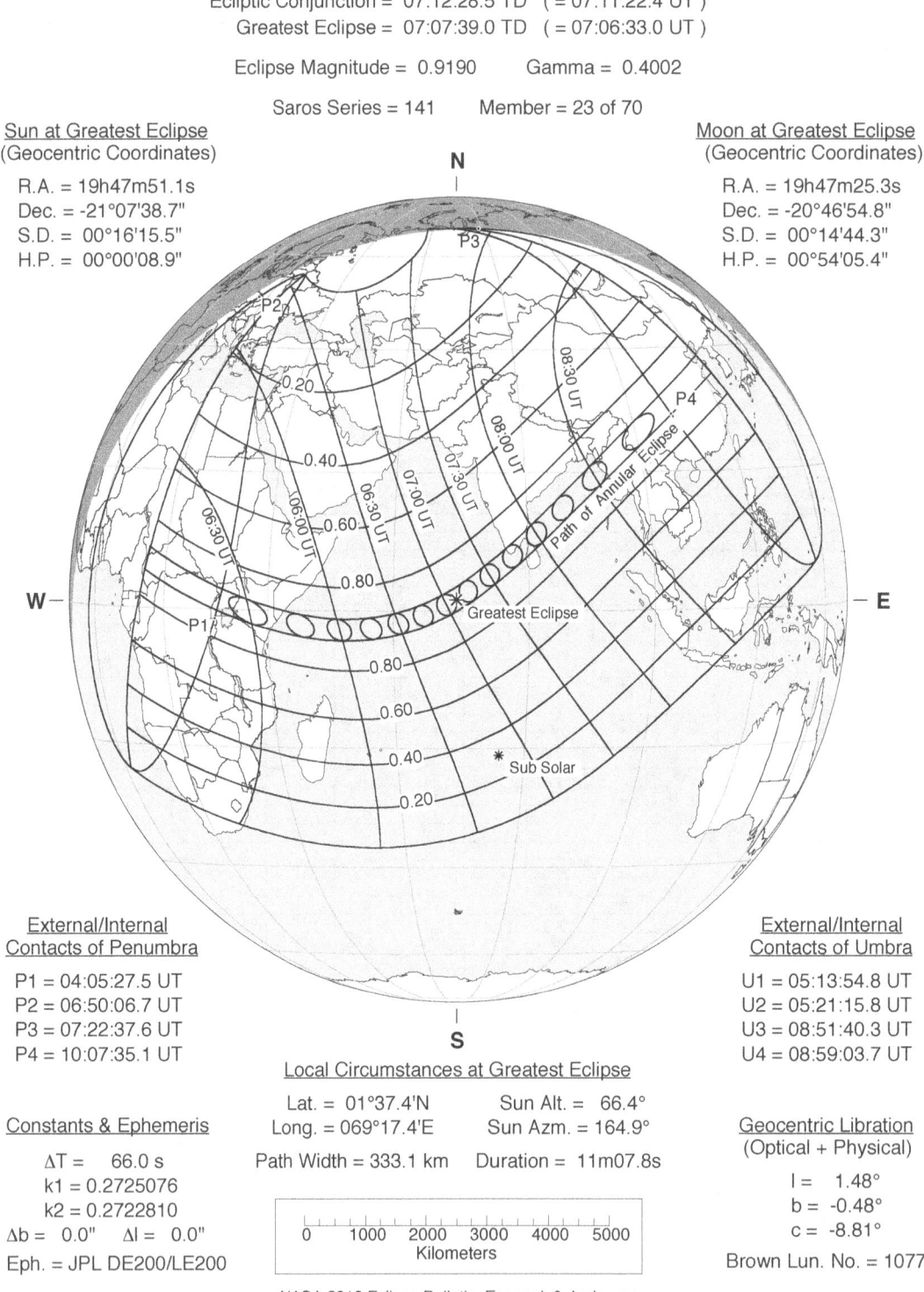

NASA 2010 Eclipse Bulletin, Espenak & Anderson

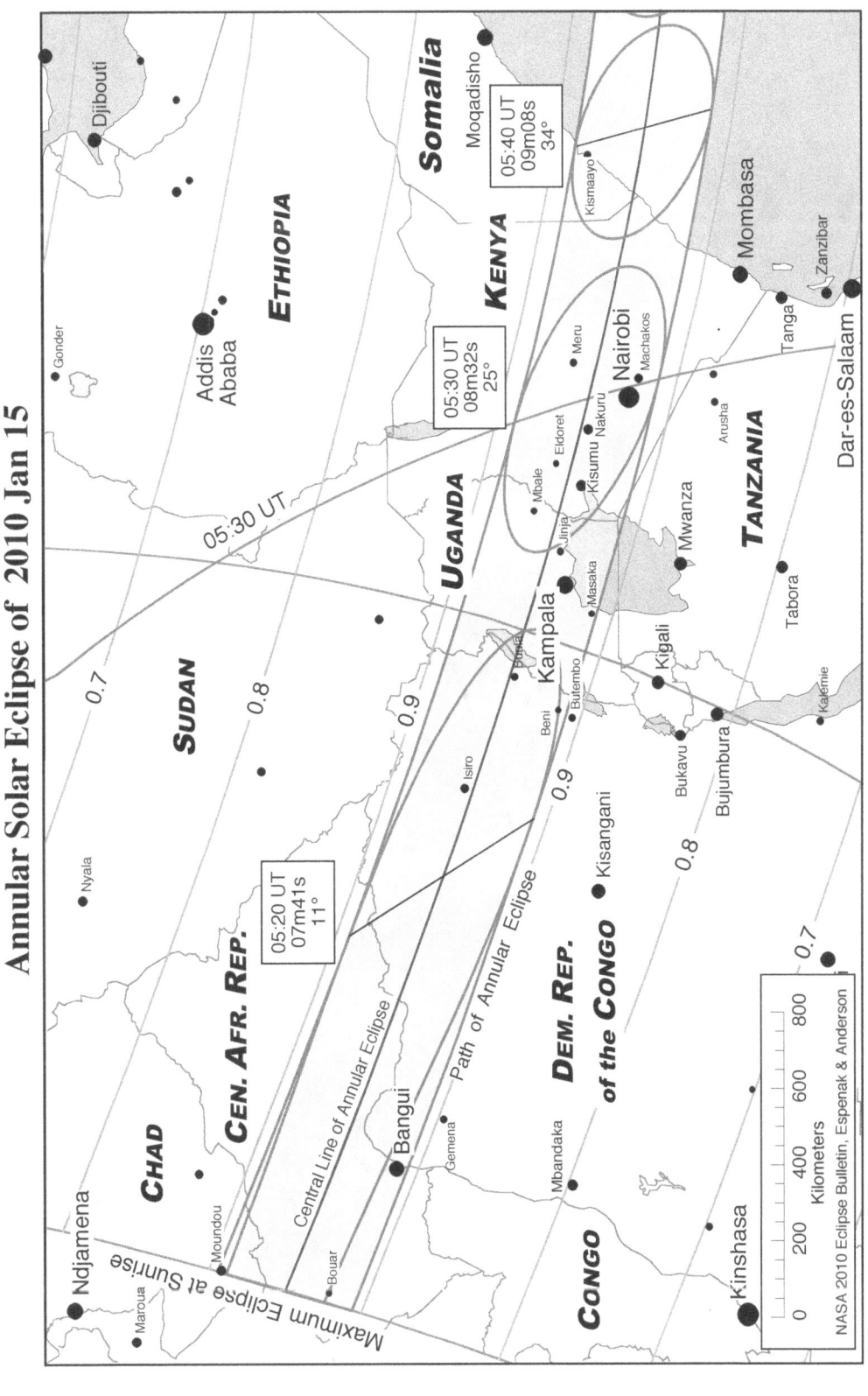

FIGURE 2.2: PATH OF THE ECLIPSE THROUGH AFRICA
Annular Solar Eclipse of 2010 Jan 15

Annular Solar Eclipse of 2010 January 15

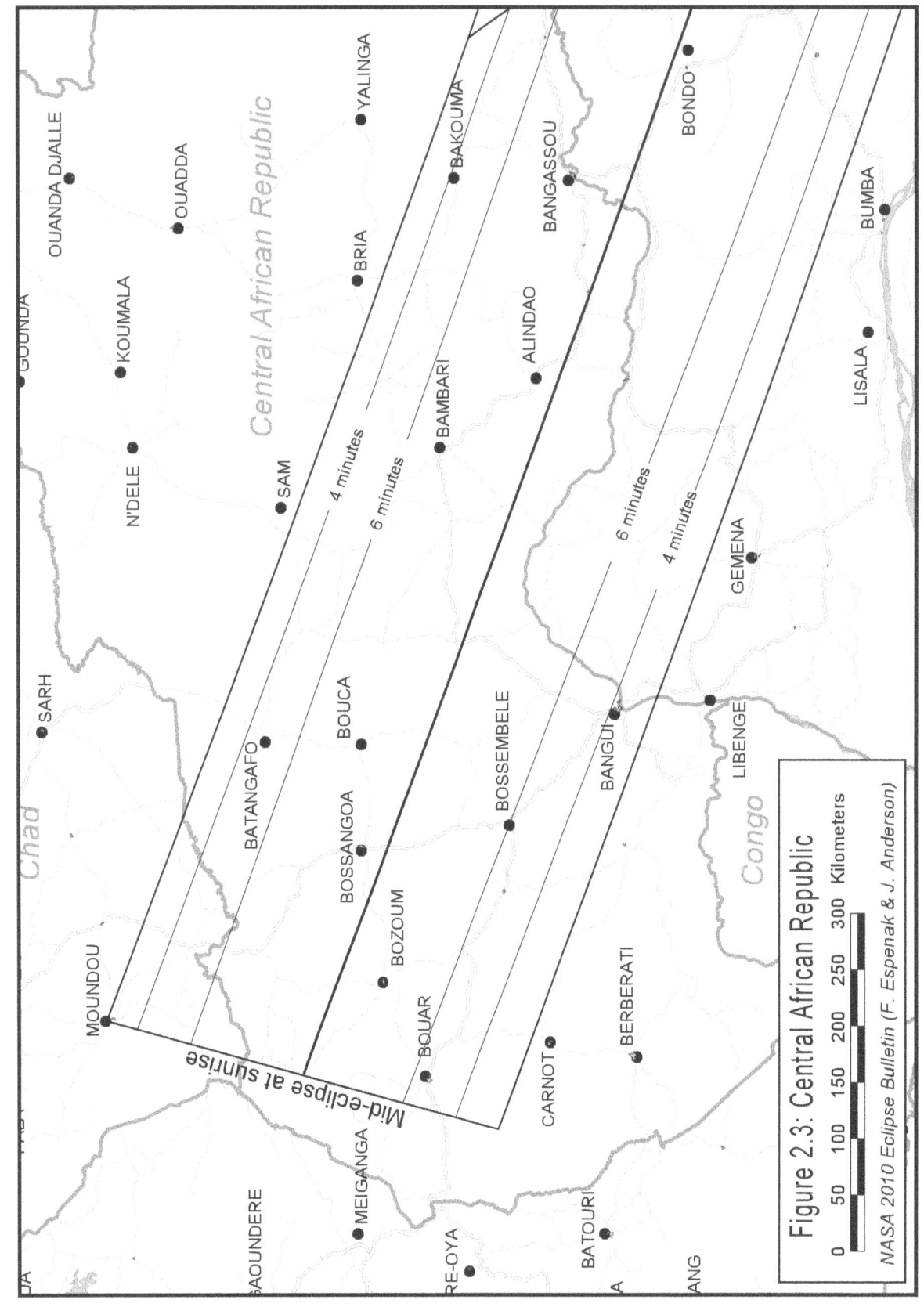

Figure 2.3: Central African Republic

Annular Solar Eclipse of 2010 January 15

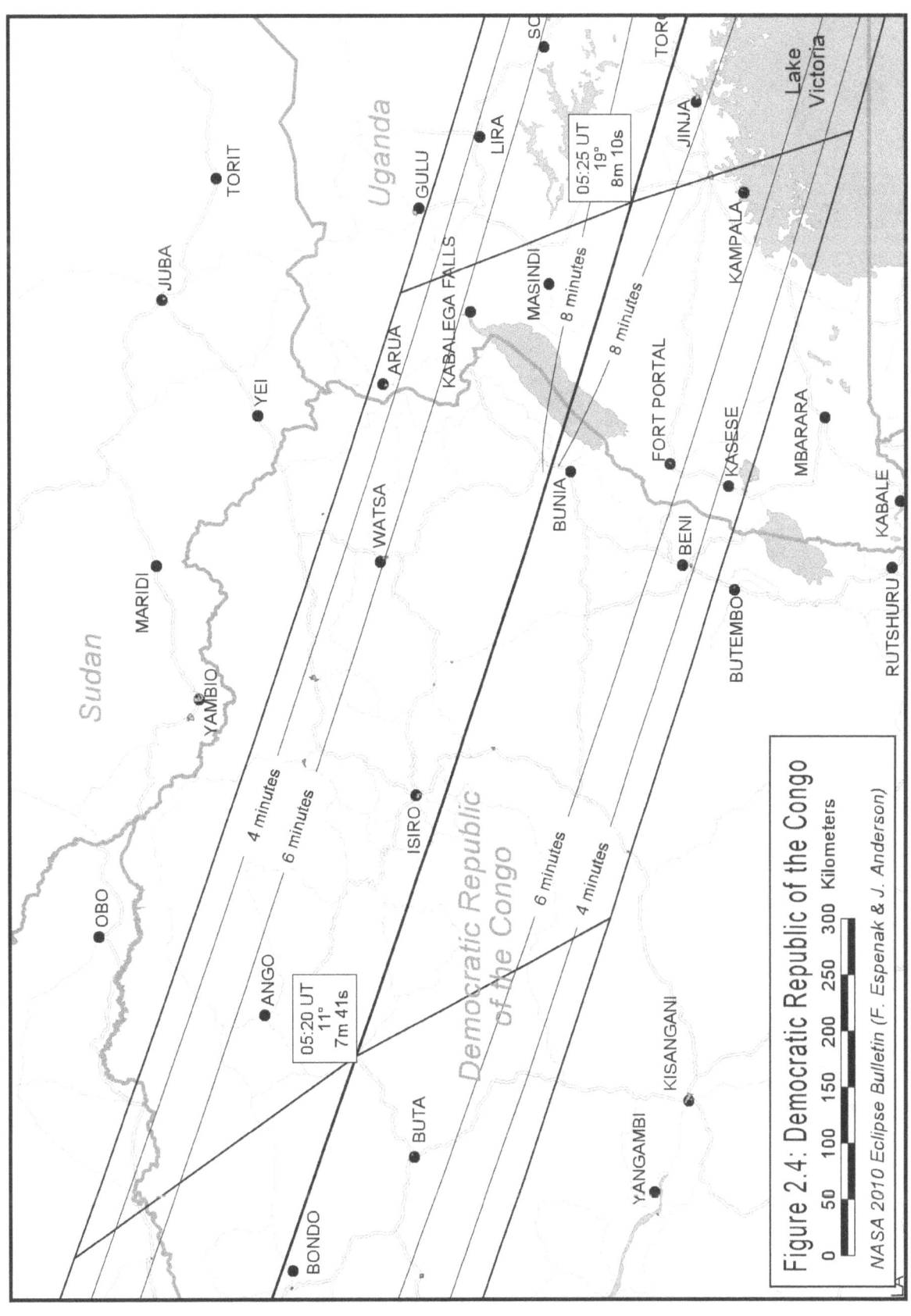

Figure 2.4: Democratic Republic of the Congo

Annular Solar Eclipse of 2010 January 15

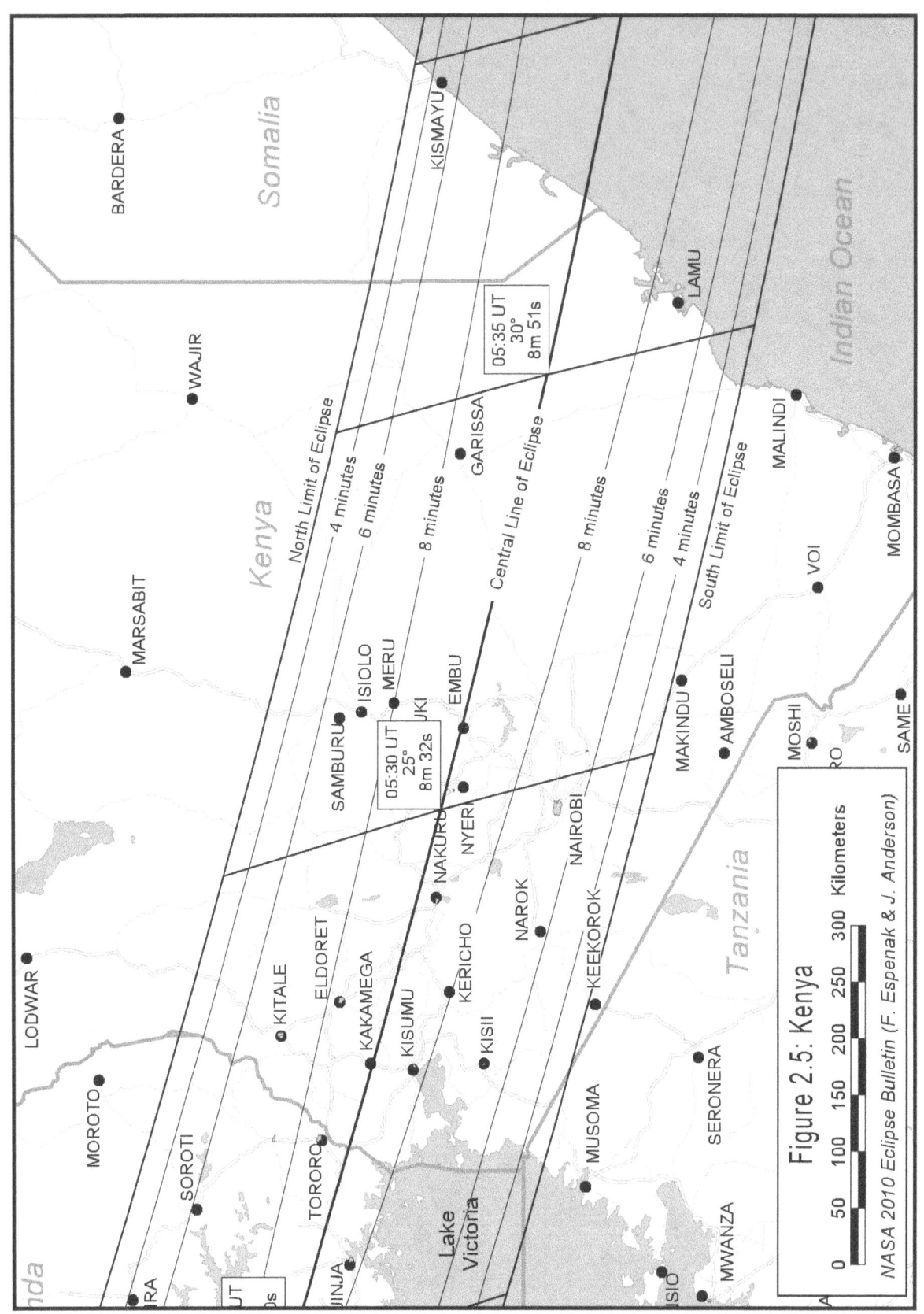

Figure 2.5: Kenya

NASA 2010 Eclipse Bulletin (F. Espenak & J. Anderson)

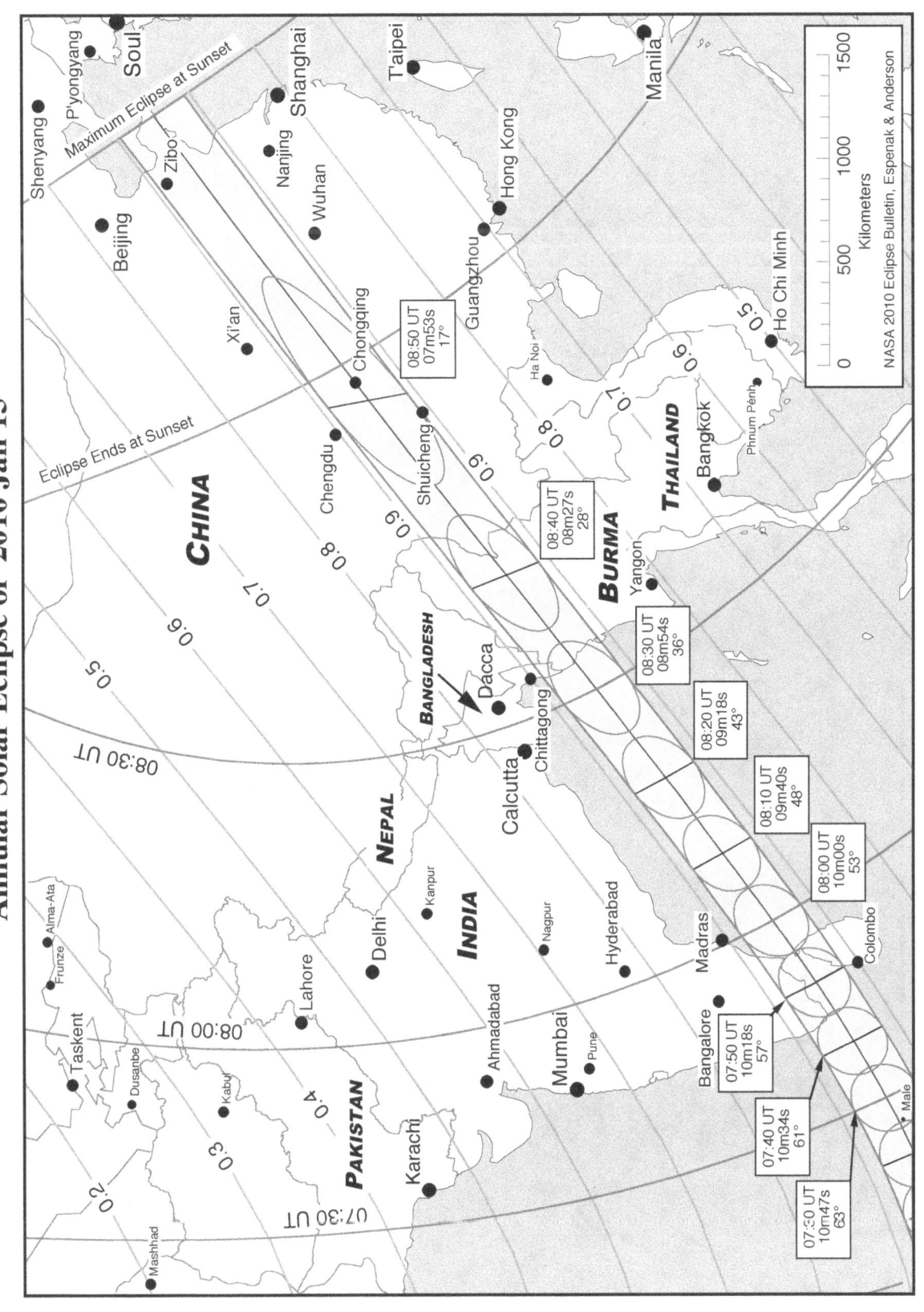

FIGURE 2.6: PATH OF THE ECLIPSE THROUGH ASIA
Annular Solar Eclipse of 2010 Jan 15

ANNULAR AND TOTAL SOLAR ECLIPSES OF 2010   31

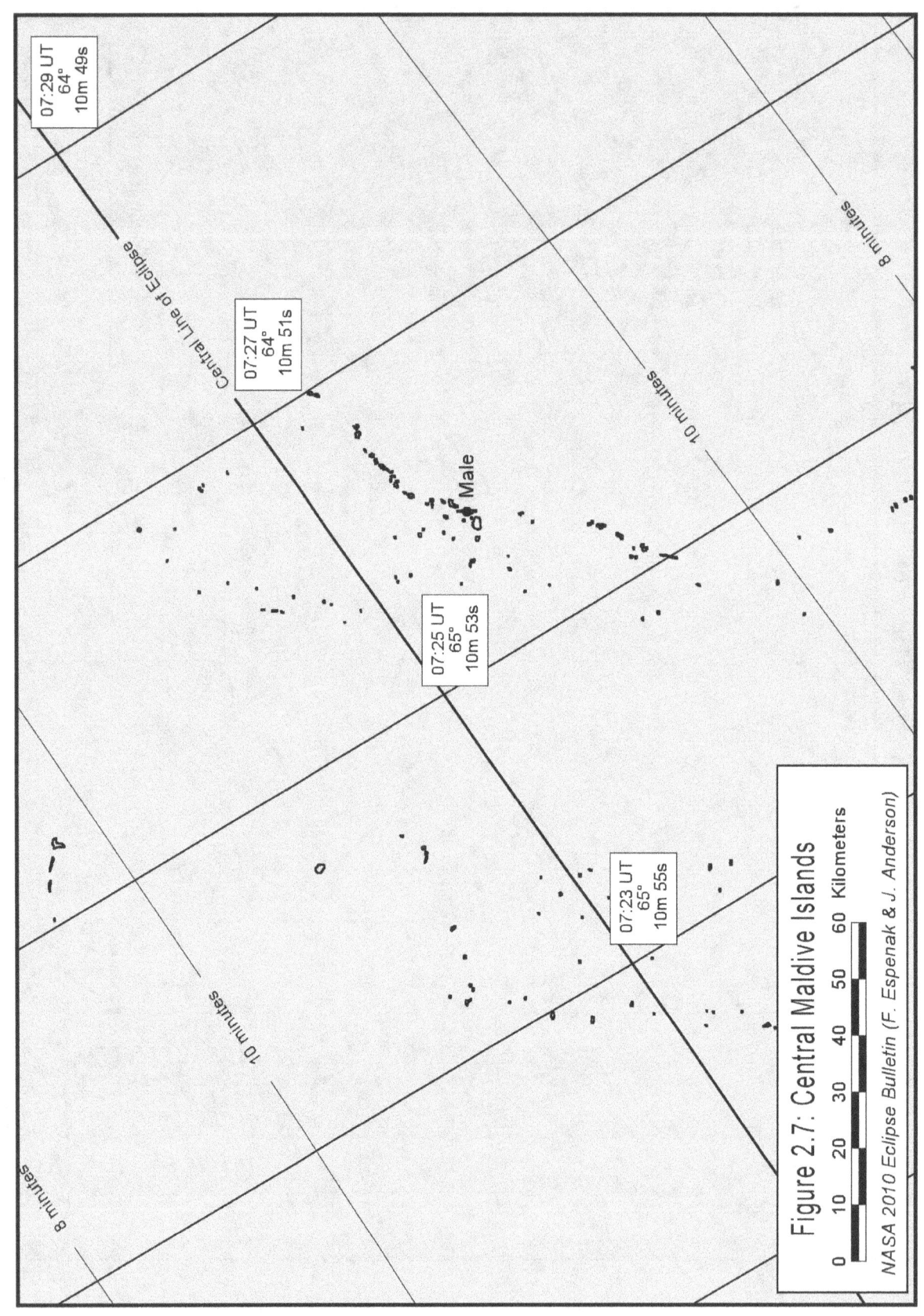

Figure 2.7: Central Maldive Islands

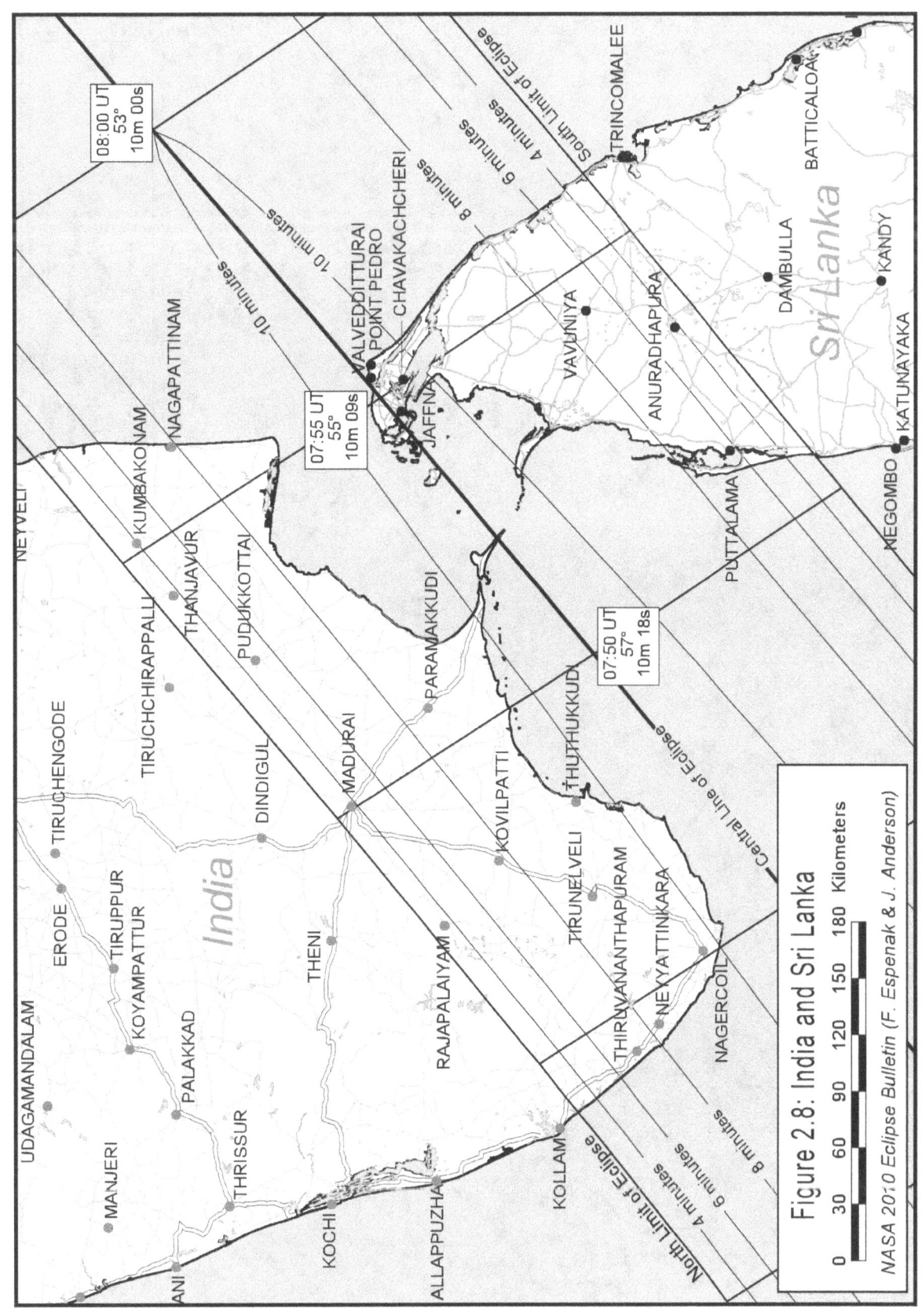

Figure 2.8: India and Sri Lanka

NASA 2010 Eclipse Bulletin (F. Espenak & J. Anderson)

Annular Solar Eclipse of 2010 January 15

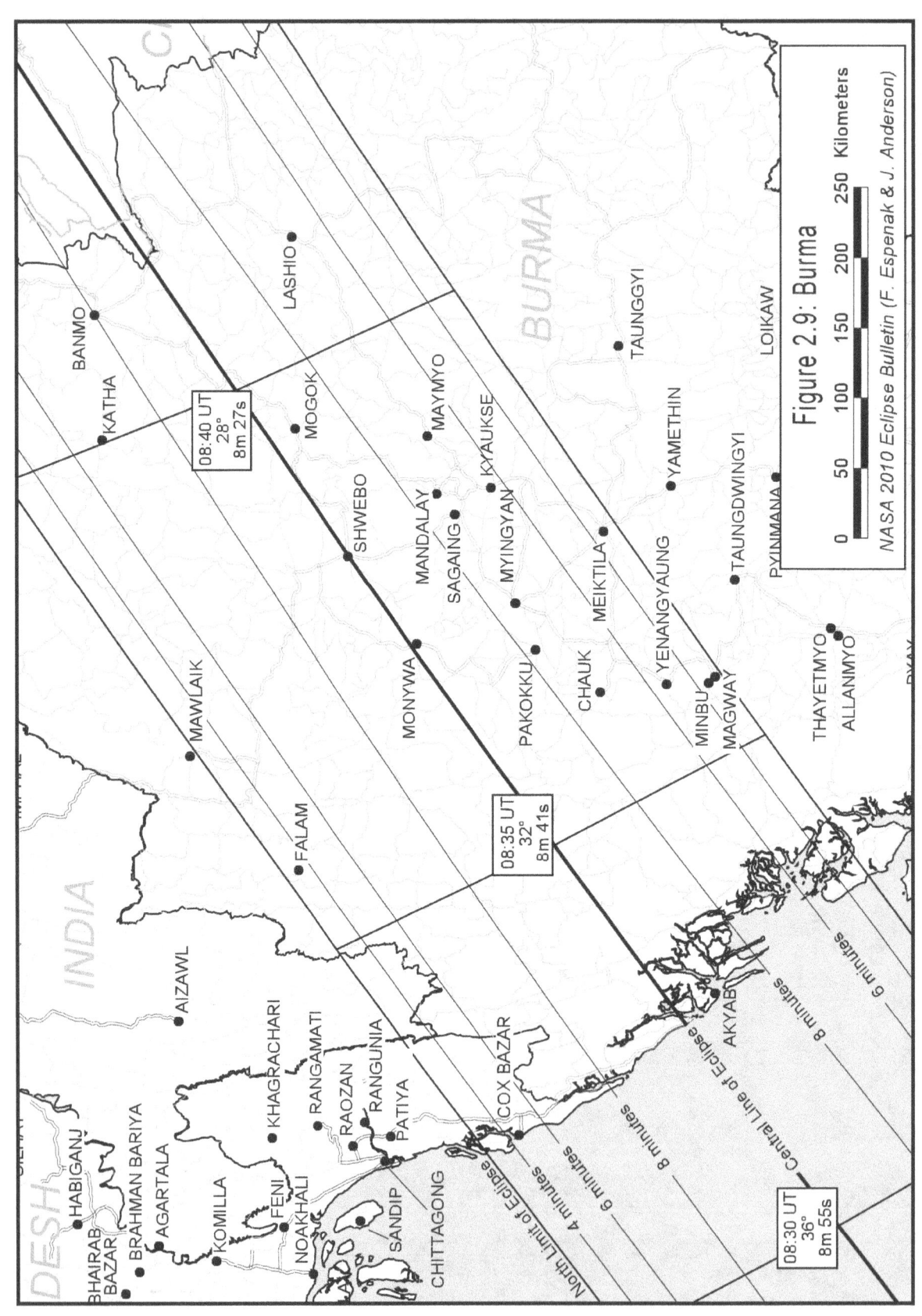

Figure 2.9: Burma

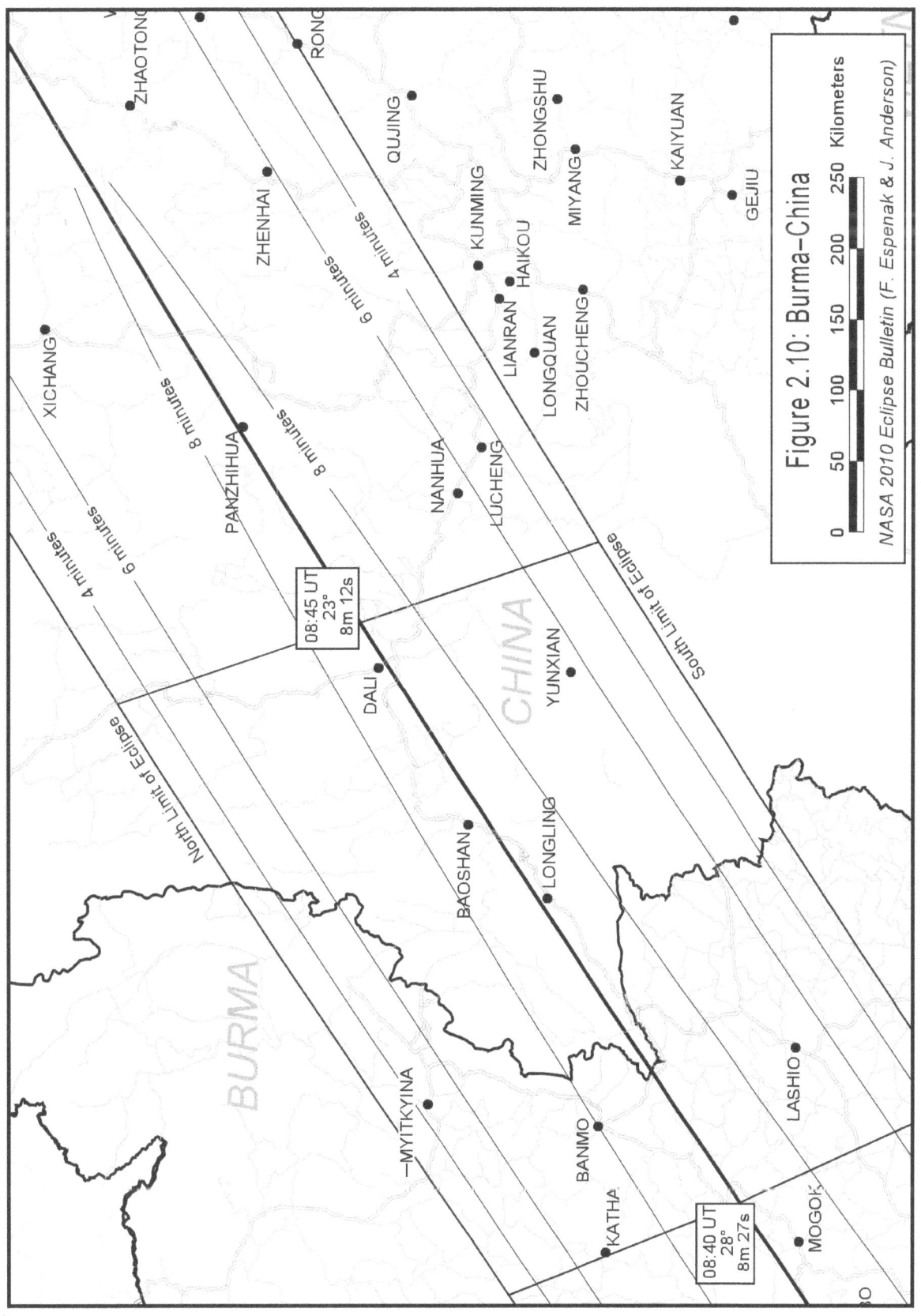
Figure 2.10: Burma–China

Annular Solar Eclipse of 2010 January 15

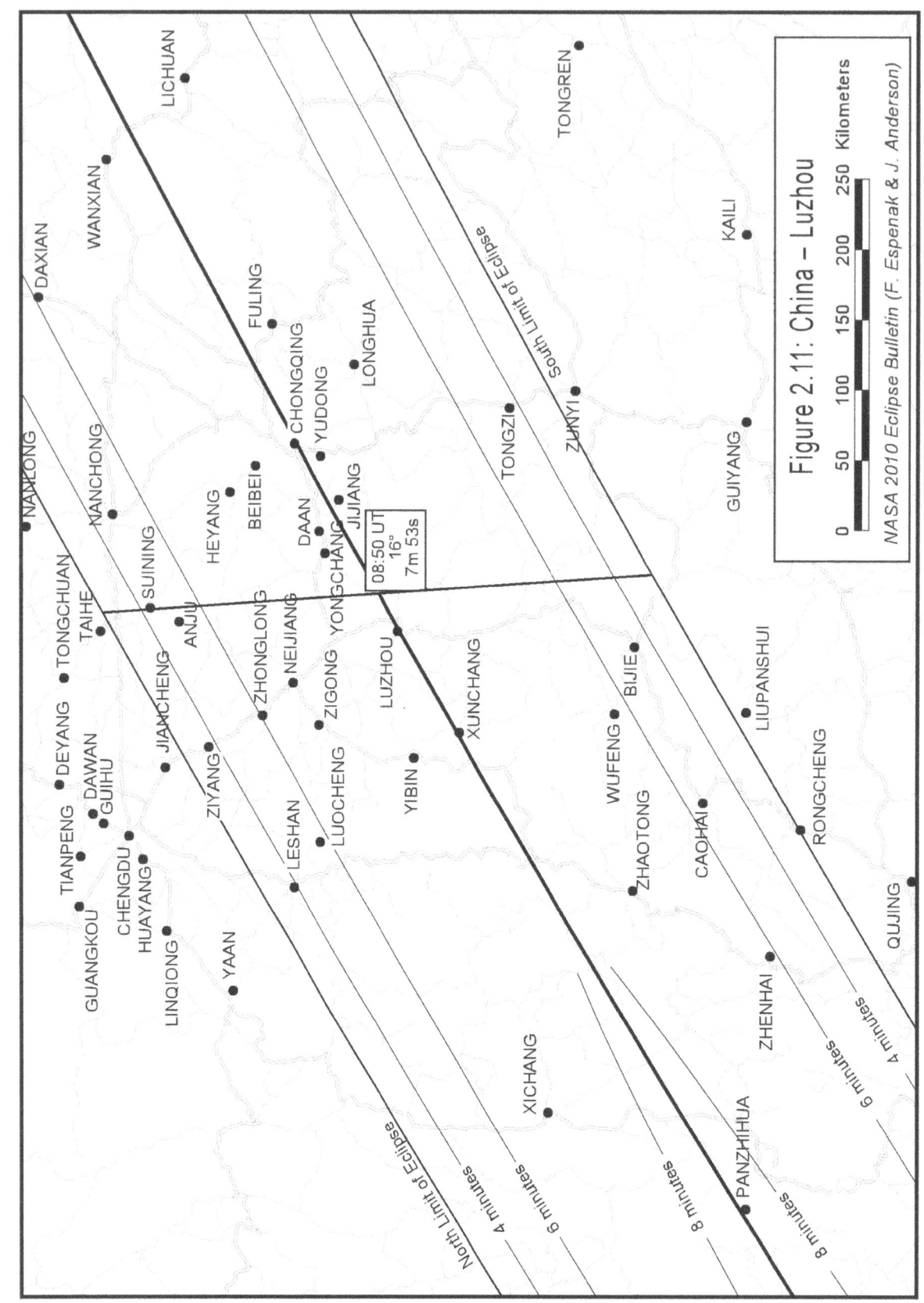

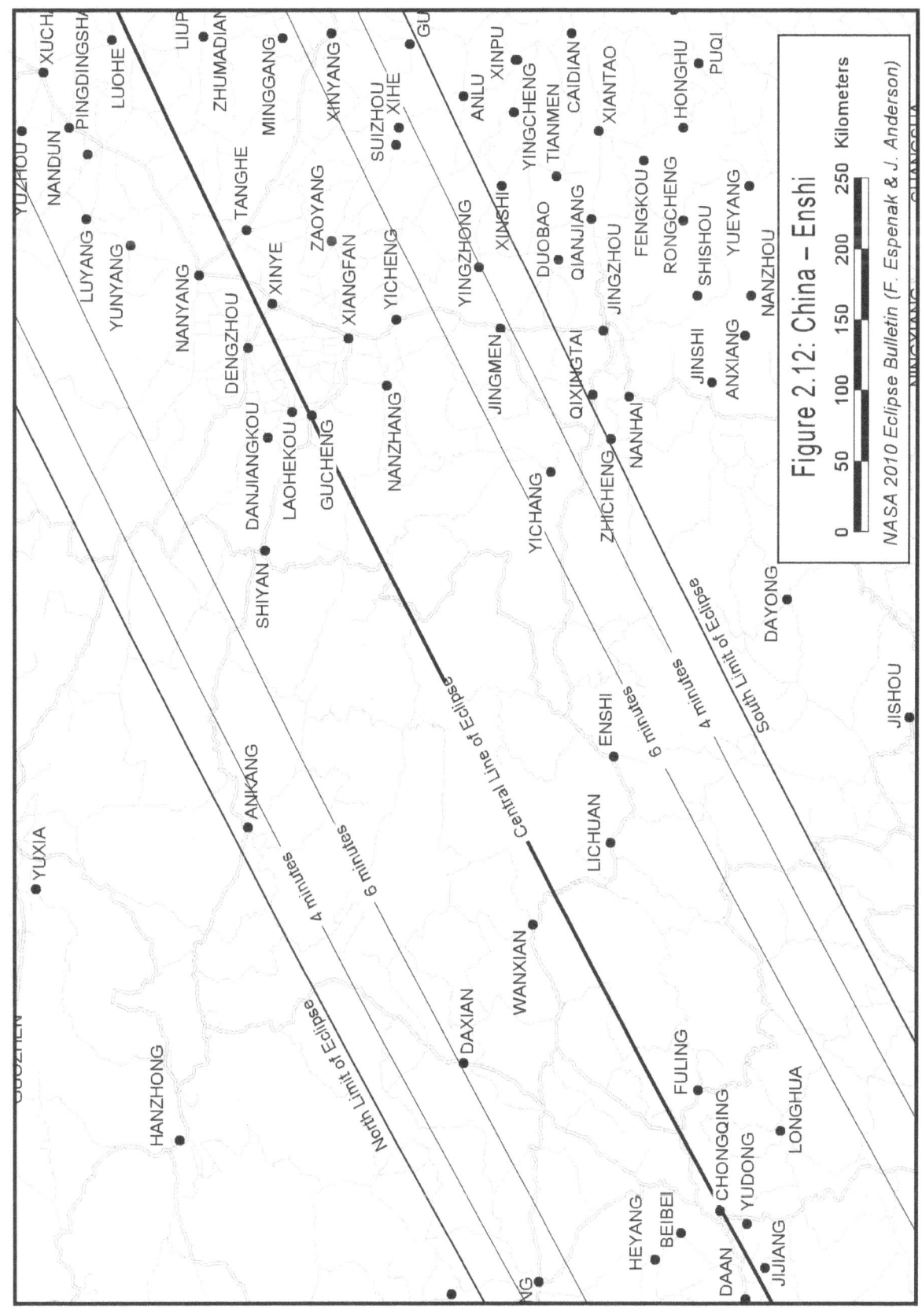

Figure 2.12: China – Enshi

Annular Solar Eclipse of 2010 January 15

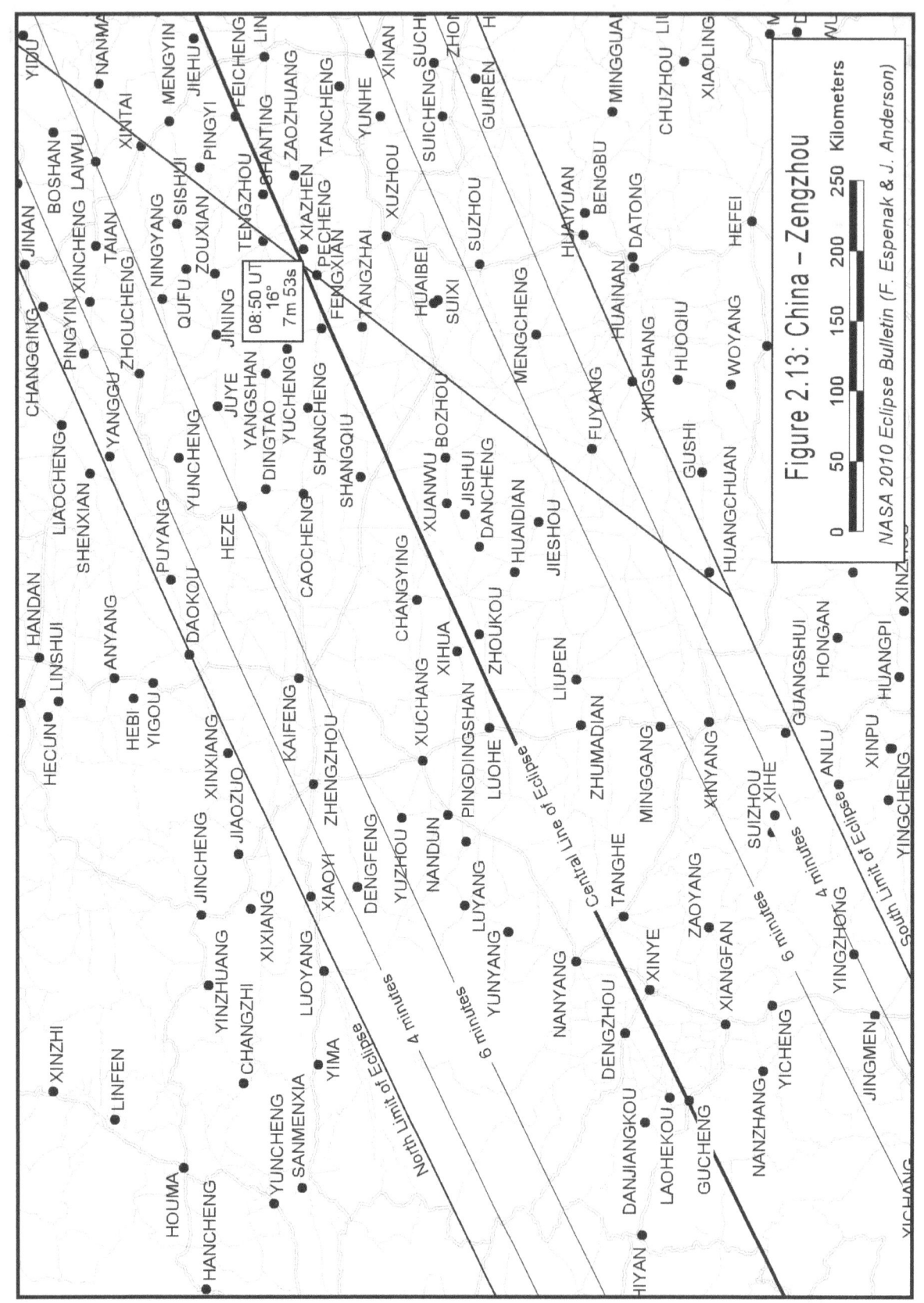

Figure 2.13: China – Zengzhou

Annular Solar Eclipse of 2010 January 15

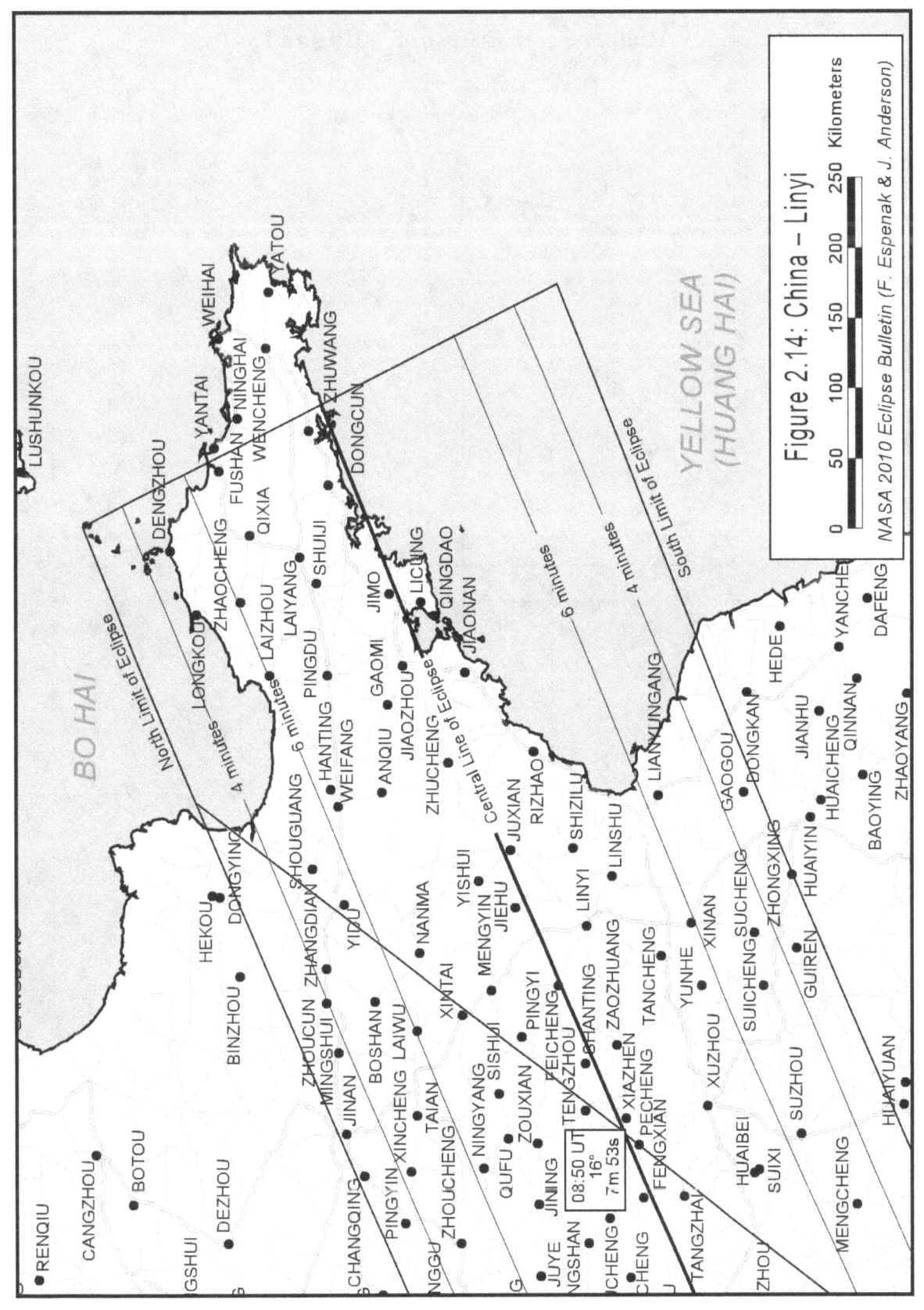

Figure 2.14: China – Linyi

ANNULAR SOLAR ECLIPSE OF 2010 JANUARY 15

FIGURE 2.15: LUNAR LIMB PROFILE FOR JANUARY 15 AT 07:00 UT

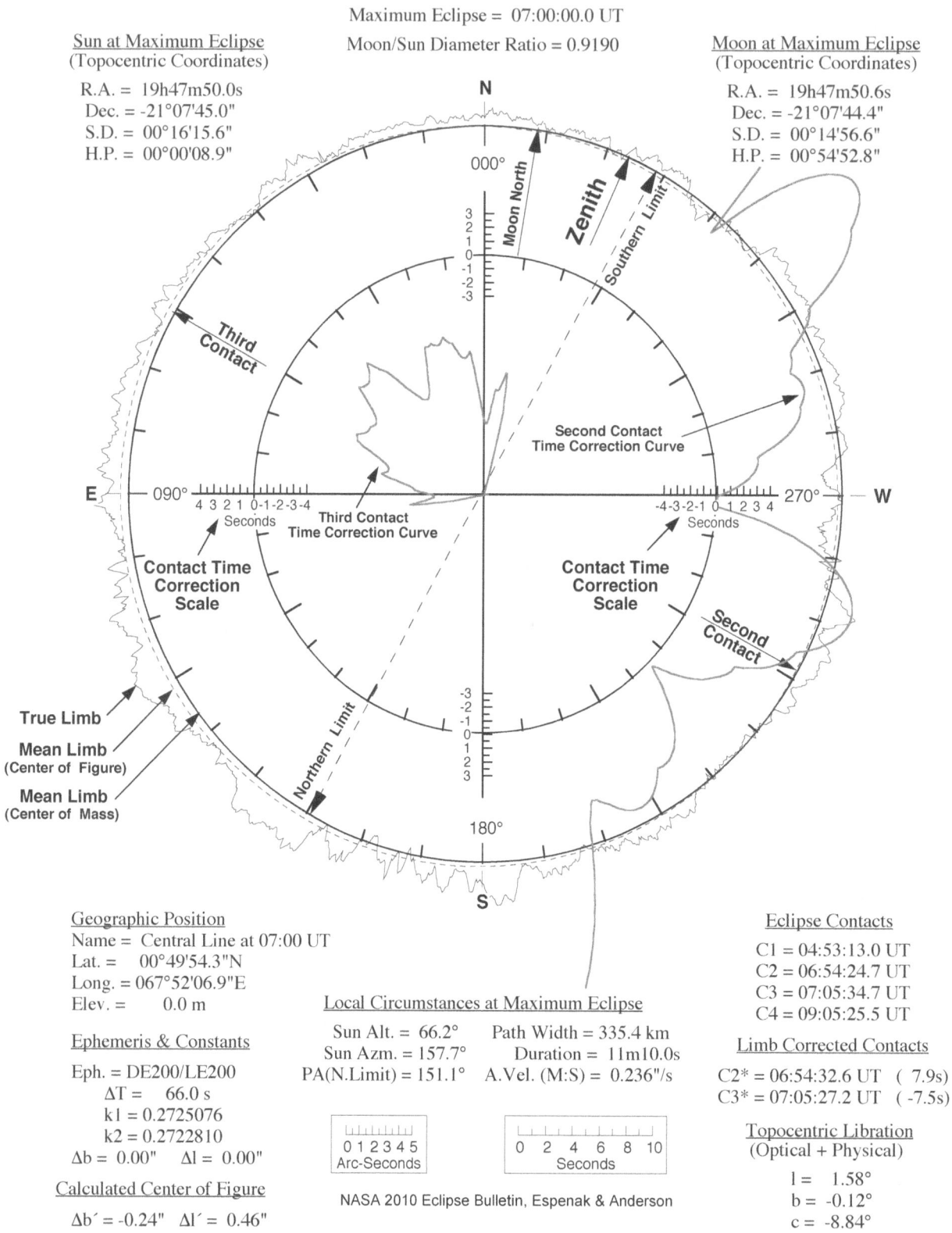

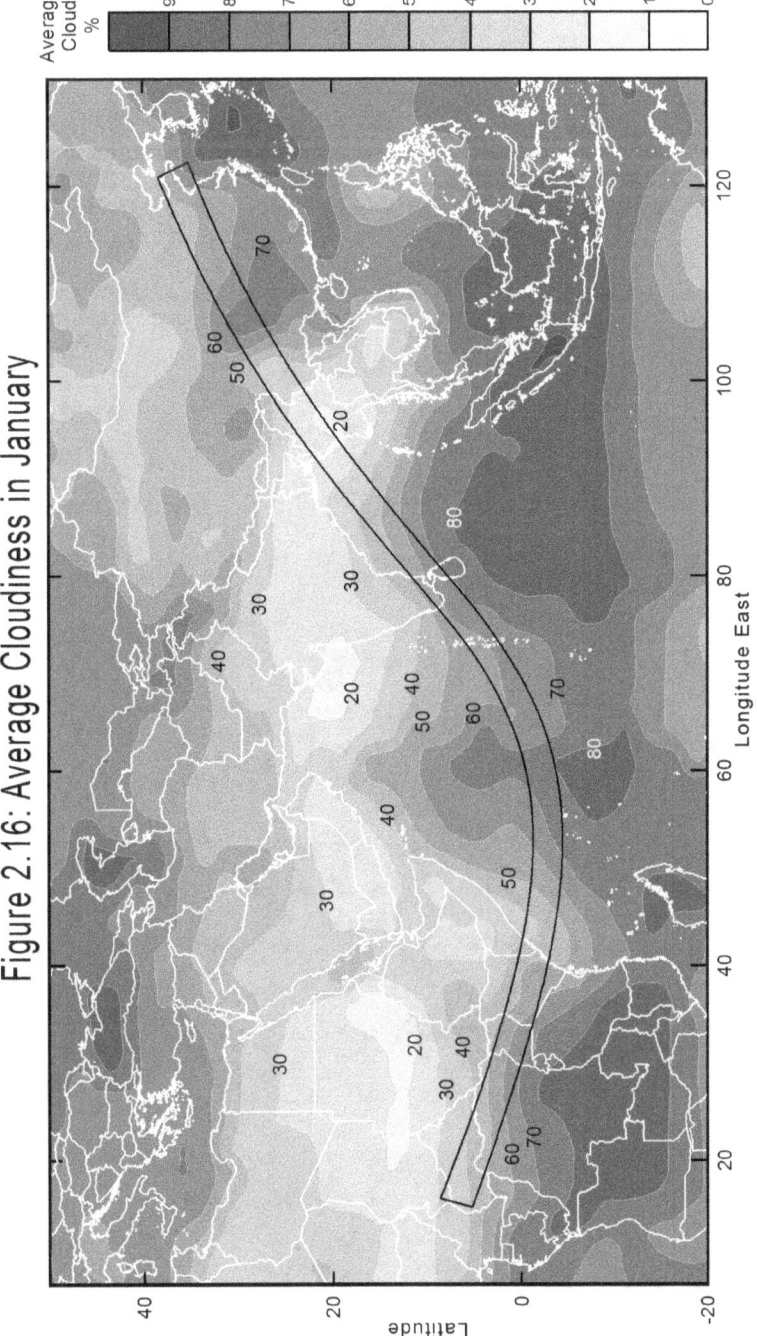

Figure 2.16: Average cloudiness in January is shown along the eclipse path as measured by a suite of orbiting weather satellites over a 23-year period. The high levels of cloudiness extending northward toward the toe of the Arabian Peninsula along 60°E longitude is a spurious effect of the data extraction. Data ISCCP.

3. Total Solar Eclipse of 2010 Jul 11

3.1 Introduction

On Sunday, 2010 July 11, a total eclipse of the Sun is visible from within a narrow corridor that traverses Earth's Southern Hemisphere (Espenak and Anderson 2006). The path of the Moon's umbral shadow crosses the South Pacific Ocean where it makes no landfall except for Mangaia (Cook Islands) and Easter Island (Isla de Pascua or Rapa Nui). The path of totality ends just after reaching southern Chile and Argentina. The Moon's penumbral shadow produces a partial eclipse that is visible from a much larger region covering the South Pacific and southern South America (Figure 3.1).

3.2 Umbral Path and Visibility

At 18:15 UT, the Moon's umbral shadow first makes contact with Earth as the path of totality begins in the South Pacific. The eclipse track is 179 km wide and the duration of totality on the central line is 2 min 42 s. Regrettably, the nearest land is either 700 km to the northwest (Tonga) or 1800 km to the southwest (New Zealand). It is most unfortunate that the dearth of solid land with the track is one of the most noteworthy characteristics of this eclipse.

As the shadow travels northeast, the path grows wider and the duration increases. At 18:19 UT, the track just misses Rarotonga—the largest and most populous of the Cook Islands—by just 25 km (Figure 3.2). Rarotonga's 14,000 inhabitants witness a very deep 0.993 magnitude partial eclipse. Two minutes later (18:21 UT), the umbra makes the first of its very few landfalls after covering 1450 km of open ocean. Mangaia (Auau Enua)—the second largest and most southerly of the Cook Islands—is only 15 km south of the central line. Its normal population of 1900 will undoubtedly grow with many new visitors there to enjoy 3 min 18 s of total eclipse. The Sun's altitude is 14°, the umbra's velocity is 2.7 km/s, and the path width is 200 km.

Leaving the Cook Islands behind, the umbra passes tantalizingly close to the Society Islands of French Polynesia (Figure 3.3). Alas, such exotic destinations as Moorea, Bora Bora, and Tahiti all lie outside the eclipse track. The southeastern coastline of Tahiti lies just 20 km beyond the northern limit and gets a 0.996 magnitude partial eclipse at 18:28 UT. In comparison, Papeete—Tahiti's capital (pop. 131,000) on the northwest coast—experiences a 0.984 magnitude partial. Because of its close proximity to the path, one or more eclipse cruises will be based out of Tahiti. While they cannot offer the rigid stability of a land-based observing site, the mobility of cruise ships can increase the possibility of avoiding clouds.

After the Society Islands, the Moon's shadow passes over a number of small atolls of the Tuamotu Archipelago (Figure 3.4), only a few of which are inhabited. Hikueru—a 10 × 15 km oval-shaped atoll—has a population of ~125 and a small territorial airport. It may be suitable for small expeditions who can expect the total phase to last 4 min 20 s. One of the more isolated atolls of the Tuamotus is Tatakoto. The 4 × 14 km wide islet has a population of ~250 and is within 20 km of the central line. The duration of totality is 4 min 35 s with the Sun 36° above the horizon (18:48 UT).

The umbra now embarks on a lonely trek with no landfall for the next 1.4 h as it races 3300 km across the South Pacific. During this period, the axis of the Moon's shadow passes closest to the center of Earth (gamma = −0.6788) as the instant of greatest eclipse is reached at 19:33:31 UT (latitude 09° 45′S, longitude 121° 53′W). The maximum duration of totality is 5 min 20 s, the Sun's altitude is 47°, the path width is 258 km, and the umbra's velocity is 0.60 km/s.

The seclusion of the lunar shadow's solitary journey is finally interrupted at 20:11 UT when it encounters Easter Island, a.k.a. Isla de Pascua or Rapa Nui, as the Polynesians call it (Figure 3.5). Renowned as one of the world's most isolated inhabited islands, Easter Island is 3,600 km west of Chile. It possesses 887 enormous monolithic statues created by the native Rapanui people hundreds of years ago. Dominated by three extinct volcanoes, the triangular-shaped island measures approximately 11 × 23 km. From the capital, Hanga Roa, totality lasts 4 min 41 s, with the Sun at an altitude of 40°. The 3,800 inhabitants of the isle are accustomed to tourism, but the eclipse is expected to bring record numbers to this remote destination.

After Easter Island, the Moon's umbra covers another 3700 km of ocean during 38 min before beginning its final landfall along the rugged shores of southern Chile at 20:49 UT (Figure 3.6). The shadow is now an extremely elongated ellipse traveling at a velocity of 8.7 km/s and accelerating. The central line duration is 02 min 57 s with the Sun 5° above the horizon. However, the desolate, fjord-lined coast of the Chilean Archipelago affords no suitable locations for eclipse observing. Quickly crossing the Andes, the shadow enters Argentina where it encounters El Calafate, a tourist village of 8000 located on the southern shore of Lake Argentino. The Sun's altitude is only 1° during the 2 min 47 s total phase, but the lake may offer an adequate line-of-sight to the eclipse hanging just above the Andes's silhouetted skyline.

The path ends 130 km southeast of El Calafate as the umbra slips off Earth's surface and returns to space at 20:52 UT. Over the course of 2 h 39 min, the umbra travels along a track approximately 11,100 km long that covers 0.48% of Earth's surface area. It will be 29 months before the next total solar eclipse occurs on 2012 Nov 13.

3.3 Maps of the Total Eclipse Path

Maps of the Jul 11 total eclipse path are given in Figures 3.1 through 3.6. Figure 3.1 is an orthographic projection map of Earth showing the path of penumbral (partial) and umbral (total) eclipse. The limits of the Moon's penumbral shadow define the region of visibility of the partial eclipse. The much narrower path of the umbral shadow defines the zone where the total eclipse is visible. For a more detailed description of Figure 3.1, see Section 1.2.

Figures 3.2 through 3.6 offer more detailed maps of the path of totality wherever it crosses land. A complete description of these figures can be found in Sections 1.3 and 1.4.

3.4 Total Eclipse Elements and Path Tables

Tables 3.1 through 3.7 give elements for the eclipse, as well as basic characteristics of the path of totality. The geocentric ephemeris for the Sun and Moon, various parameters, constants, and the Besselian elements (polynomial form) are found in Table 3.1. All external and internal contacts of penumbral and umbral shadows with Earth are listed in Table 3.2. They include TDT and geodetic coordinates with and without corrections for ΔT.

The path of the umbral shadow is delineated at 5 min intervals (in Universal Time) in Table 3.3. Coordinates of the northern limit, the southern limit, and the central line are listed along with the Sun's altitude, path width, and central line duration of totality. Table 3.4 presents a physical ephemeris for the umbral shadow and includes the topocentric ratio of the Moon and Sun's apparent diameters, the eclipse obscuration, the path width, the dimensions of the umbral shadow, and its ground velocity.

Table 3.5 gives the local circumstances for each central line position listed in Tables 3.3 and 3.4. Table 3.6 presents topocentric values from the central path for the Moon's horizontal parallax, semi-diameter, relative angular velocity with respect to the Sun, and libration in longitude. In addition, corrections to the path limits due to the lunar limb profile are listed. A detailed description of these tables can be found in Section 1.5.

3.5 Total Eclipse Local Circumstances Tables

Local circumstances for a number of cities, islands, and places in the Pacific Ocean and South America are presented in Table 3.7. The table gives the local circumstances at each contact and at maximum eclipse for every location. The coordinates are listed along with the location's elevation (in meters) above sea level. The Universal Time of each contact is given to a tenth of a second, along with position angles **P** and **V** and the altitude of the Sun. Two additional columns are included if the location lies within the path of totality. The "umbral depth" is a relative measure of a location's position with respect to the central line and path limits. The last column gives the duration of totality. For a more information about these tables, see Section 1.6.

3.6 Total Eclipse Lunar Limb Profile

Along the 2010 total eclipse path, the Moon's topocentric libration (physical plus optical) in longitude ranges from $l=-2.5°$ to $l=-3.8°$; thus, a limb profile with the appropriate libration is required in any detailed analysis of contact times, central durations, etc. A profile with an intermediate value, however, is useful for planning purposes and may even be adequate for most applications. The lunar limb profile presented in Figure 3.7 includes corrections for center of mass and ellipticity (Morrison and Appleby 1981). It is generated for 19:30 UT, near the time of greatest eclipse. The umbral shadow is then located in the Pacific Ocean at latitude 19° 18.6′S and longitude 121° 59.4′W. The Moon's topocentric libration is $l=-3.13°$, and the topocentric semi-diameters of the Sun and Moon are 944.0 and 998.7 arcsec, respectively. The Moon's angular velocity with respect to the Sun is 0.342 arcsec/s.

The times of the four eclipse contacts from this location appear to the lower right in Figure 3.7. The limb-corrected times of second and third contacts are listed with the applied correction to the center of mass prediction. The time correction curves can be used for estimating corrections to the times of second and third contacts as a function of the position angle of the contact. More information on this topic and a detailed description of the limb profile figure can be found in Section 1.8.

3.7 Weather Prospects for the Total Eclipse

3.7.1 Weather Overview

The 2010 total eclipse comes at the depths of the Southern Hemisphere winter, ordinarily a time of frequent storms and alternating high- and low-pressure systems that bring a lot of changeable weather, winds, and cloudiness. Fortunately, latitude comes to the rescue for the first half of the shadow track, as it travels in and north of the belt of high-pressure anticyclones that girdle Earth at about 30°S (Figure 3.8). This high-pressure belt is a region where the air descends from higher levels in the atmosphere, warming and drying by adiabatic compression. It is a zone of mostly sunny skies and pleasant temperatures, akin to the Caribbean in the Northern Hemisphere winter, but it is not without its temperamental weather. Cold fronts from storms in the "Roaring Forties"—latitudes between 40°S and 60°S—are able to move into anticyclonic barrier, bringing showery weather and cloudy skies to the eclipse path when they do. Beyond Easter Island, the eclipse track dips into the Roaring Forties, and cloudiness—at least over the eastern Pacific and the coast of Chile—becomes much heavier.

In addition to passing cold fronts and the impact of the Roaring Forties, there is a semi-permanent feature of southern meteorology known as the South Pacific Convergence Zone (SPCZ). The SPCZ is a band of low-level wind convergence lying over the warmest waters of the southwest Pacific and so, like the ITCZ along the equator, is a region of frequent showers and thundershowers along with the associated cloudiness. In July, alas, the SPCZ lies at the northern limit of its annual range, stretching from the Solomon Islands near New Guinea, across Samoa and the Cook Islands. In recent years, perhaps in response to global climate changes, the SPCZ has tended to move north and east, to a position that more directly affects the eclipse track. To avoid the influences of the low-latitude storms and the thunderstorms of the SPCZ, eclipse watchers must head to the north and east—to the extremities of French Polynesia or beyond (Figure 3.9).

In spite of the weather factors that promote cloudiness along the eastern and western extremities of the eclipse path, the western Pacific is actually in the midst of its seasonal dry spell during July. Closer to the South American coast, just the opposite is true, though at Easter Island, the difference in precipitation between the wet and dry seasons is less pronounced than at Tahiti and the Cook Islands (Figure 3.10). At the end of the path in Argentina, where winter influences would be expected to bring the most difficult conditions, the Andes Mountains act as a very effective barrier to the Pacific storms and the weather is quite promising instead.

Islands in the Cook Islands and French Polynesia are either mountainous volcanic peaks (Tahiti, Mangaia, and also Easter Island) or low, flat atolls (Tuamotus). The latter are too small and low to affect the flow from the sea, so the weather observations from those sites reflect the conditions on the water. On the other hand, the mountainous islands impose a considerable orographic modification on wind, cloud, and precipitation—generally to increase cloud and rainfall and divert the winds. The humid tropical air is always ready to form clouds if lifted by any of several processes. Large islands are darker than the sea, and warm more readily in sunlight. Warm air, being buoyant, rises upward, forming clouds at some small distance above the surface. Winds blowing onto the land are compelled to rise as they encounter the mountainous topography, adding to the impact of the solar heating, and usually cloaking the mountain ridges and peaks with a cap of cloud, especially in the afternoon hours.

The reverse process occurs at night in the case of solar heating, and on the lee side of the terrain in the case of orographic lifting. Winds blow downslope on crossing the highest point of the terrain, so clouds dissipate and rain ends. The whole process is complicated by the complexities of the topography, but in general, the lee side of the mountains on Mangaia and Easter Island will have a slightly greater tendency to sunny weather. The degree of impact will depend on the height and lie of the terrain, and on both islands the cloud-producing processes will dominate those that dry out the air.

3.7.2 Cook Islands

Mangaia, the only Cook Island within the eclipse path (Figure 3.2), has a latitude that puts it securely within the influence of the SPCZ though the Convergence Zone has a mixed personality, sometimes quiet and barely evident, other times especially active and full of convection and rain. Cold fronts, migrating northward from lower latitudes, reinforce the SPCZ or attend the islands with their own independent weather. It is easy to be pessimistic about the weather prospects, but the climate statistics for Mangaia give reason for some optimism: July is the driest month at nearby Rarotonga (Figure 3.10), with an average of about 100 mm of rainfall and Mangaia follows the same pattern. Mangaia's cloudiness is similar to that of Rarotonga (Table 3.8) with an average cloud cover of 64% calculated from the observed frequency of the various cloud categories. Rarotonga reports an average sunshine amount of 52% and it is probably only slightly less than this at Mangaia.

Mangaia Island is 9 km in diameter, rugged, with a modest 170 m peak in its interior. It is mostly tree-covered in its interior and so combines the cloud-producing features of low albedo and a rising topography, although this forcing is modest compared to Tahiti. On most days when the SPCZ is weak or distant and the skies are sunny, the afternoon convective clouds are small and confined to the interior. Clouds such as these will dissipate quickly in the cooling that accompanies an eclipse. When more organized weather visits, the small dimensions of the island and limited terrain are unlikely to have much influence, either to reinforce the rainfall on the windward side or to dissipate the clouds to leeward. Whatever influence the island can muster will be confined to the lowest cloud levels. There is no prevailing wind at Mangaia, but the stronger weather systems tend to come with easterlies and southeasterlies.

On quiet days, rains on Mangaia tend to come in the afternoon after a sunny morning, and may be quite heavy for a brief time. Daylong rains are more unusual, but do occur from time to time. This diurnal pattern favors the eclipse, which occurs in the morning hours, before maximum heating and maximum cloudiness. Winds blowing against Mangaia (on the windy days—one-third of the wind observations are calm) may cause the formation of an arc of cloudiness offshore where the winds converge and are diverted to flow around the island. These arc clouds (much like the bow wave of a boat) will likely remain offshore during an eclipse.

3.7.3 Tahiti and French Polynesia

At Tahiti, July is the second-driest month (Figure 3.10) and at Hereheretue, in the Tuamotu Islands, July is the driest. While this pattern is similar to that at Rarotonga, the amount of rain in July is about half that in the more southeasterly Cook Islands. The drier weather is reflected in the cloud cover statistics, with average cloudiness dropping to between 44% and 53% across much of Polynesia, a figure 10% to 20% less than in the Cook Islands, in large part due to the reduced influence of the SPCZ. Sunshine statistics are also generous, though somewhat erratic, with very encouraging measurements of 65% to 70% of the maximum possible in most of the islands.

Periods of bad weather are often associated with the passage of cold fronts that arrive from the southwest, sometimes lingering for several days. The stronger fronts have a tendency to stall near the islands and the eclipse track. Even though Tahiti is in the midst of its dry season, a persistent frontal band can drop large amounts of rain for several days in a row. Should a cold front occur on eclipse day, the only escape would be to sail out from under it, most likely by heading eastward down the path.

With only a limited number of places to stay on Mangaia and the atolls of French Polynesia, most southwest Pacific observers will choose to watch this eclipse from shipboard. From a climatological perspective, ships should place themselves as far eastward along the track as schedule permits. This puts the SPCZ well behind and increases the probability that temperate-zone cold fronts will be left behind (Figure 3.9). Cloud systems tend to become smaller and more disorganized in the more northerly latitudes, and thus easier to avoid when eclipse day arrives. Closer to Tahiti, the SPCZ cloudiness will have to be watched cautiously, though the island typically marks the easternmost extent of its influence.

In satellite imagery, high- and mid-level clouds associated with the SPCZ tend to move from west to east in the upper level flow. Low-level clouds usually move in the opposite direction, but are much more variable and cannot be counted on from one day to the next; frequently they just hang around without seeming to go anywhere. When strong highs pass to the south of the island, the pressure gradient is compressed and stronger than normal easterly trade winds—known as a Mara'umu—can bring winds of 50 km/h and 3 m wave heights in the seas, sometimes lingering for days or even over several weeks. These enhanced trade winds bring heavy rains to the windward side of Tahiti, but such orographic effects will not be a factor along the eclipse track as there are no islands with a significant topography under the lunar shadow's path in French Polynesia.

Frontal clouds will come up from the south or southwest, usually in the form of bands that are 100 or 200 km wide. On top of this complicated pattern is the tendency for clouds—especially low-level clouds—to form and dissipate over one- to three-hour periods, making prediction from satellite imagery very difficult. "Chasing" an opening in the clouds may be a frustrating experience so positioning the ship where climatology is most favorable at the start will make eclipse-day planning a less hectic event. Positions to the northeast along the path will also increase the eclipse duration. Cyclones should not be a problem, as the hurricane season runs from November to March.

When the SPCZ keeps to the south (its more usual position) and cold fronts are not in the area, westward-moving cloud clusters known as "easterly waves" may be the only weather feature to watch for. Easterly waves are more-or-less circular areas of convection with varying dimensions, up to 200 km or thereabouts. They may bring heavy overcast or scattered thundershowers, but are readily seen and predicted in satellite images.

A careful watch on the satellite images will show "zones" of descending air where both high and low-level clouds tend to disappear. These zones will not be very distinct, and they will not be completely free of cloud, but once identified, can be counted on for favorable circumstances for a half-day or longer. From the ship's deck, such areas will have smaller convective clouds (primarily shallow cumulus), and thinner high-level cirrus.

A ship's mobility will increase the chances of seeing the eclipse by an estimated 5%, limited, in large part, because cloud patterns are not easily predicted. The biggest advantage to be given to a shipboard site is the ability to move east of the Tuamotu Islands to tap the best climatology along the track. For those eclipse watchers who are determined to settle on land, the Tuamotus offer a few places with airports, including Hao, Hikueru, Tatakoto, and Anaa. Other islands will have to be reached by boat, a prospect that greatly extends the travel time.

3.7.4 Tropical Cyclones

In the southern Pacific, the tropical cyclone season runs from November to April. For regions along the eclipse path, the frequency is relatively low, with about nine storms per year on average across the whole basin east of Australia. While statistics are somewhat poor, the Cook Islands near Rarotonga (including Mangaia) experience one tropical storm every seven years, while in Tahiti they are about half that rate. In El Niño years, cyclones tend to be widespread between 10° and 30° S latitude, from Australia to 130°W, which pretty much covers the whole track through the Cook Islands and Polynesia. In La Niña years, cyclones tend to be fewer in number, forming and traveling much closer to the Australian coast. In any event, the possibility of a tropical cyclone is virtually nil during July.

3.7.5 Easter Island

Easter Island lies on the south side of the anticyclonic belt that circles Earth at 30° S latitude, and as a consequence, is much more exposed to the influence of the westerlies and storms in the Roaring Forties. In July, Easter Island is in its wet winter season (Figure 3.10) and sunshine is at a premium. Still, it is an exotic destination, and the sunshine statistics show a percentage of the maximum possible (50%) that is similar to that at Rarotonga (and Mangaia) in the Cook Islands (Table 3.8). The prospect of stunning photographs of the eclipse over the Moai has tremendous appeal.

Easter Island has three large volcanoes and a number of smaller ones, and the cloud on the mountaintops is a persistent feature of the winter weather. The weather is extremely changeable when it is inclined to be cloudy, and there is no advantage in chasing from one site to the other at the last minute to find a sunny haven. There is a strong convective element to the cloud types, even when large weather systems reach the island, and because of this, clouds can form and dissipate within minutes. On sunny days, clouds will tend to form in the afternoon, but will dissipate as the eclipse approaches.

Given the nature of the cloudiness described above, there are still a few tricks to help pick a successful eclipse site. Do not go uphill unless the day is spectacularly sunny. Especially, do not locate on the upwind side of a volcanic hill. Coastal sites exposed to the wind may have a little less cloudiness if the wind is not too strong, as the cooler air from the sea will suppress the immediate formation of cloud as it reaches land. Sites in the lee of the larger volcanoes may be a little sunnier if the weather is not too thick, but usually, the clouds will form on the slopes and blow downwind; the flanks of the larger peaks may offer safer sites. There is no strong prevailing wind—they can come from any quarter according to the weather of the day.

Given all of the complexities of the wind and weather, the south coast seems like the safest bet, perhaps at Tongariki where the Moai offer great visual appeal. With easterly or westerly winds, the village at Hanga Roa is promising, but southerlies or northerlies will carry clouds from the peaks of Terrevaka or Rana Kao onto the town. Northerlies at Tongariki will have a slight downslope flow, which tends to dry the air out a bit, but the volcano Pakaiki lies just to the east and flow from its peak will have to be watched carefully. The beach at Anakena is promising under a northerly onshore flow, and may be one of the best sites for a large group because of the facilities available there.

Unless there is a large and active weather system over the island on eclipse day, there will certainly be mixtures of sun and cloud that will make site selection difficult. If mobility is an option, eclipse watchers should wait until the last possible moment to assess the character of the cloud and wind before picking a viewing site.

3.7.6 South America

The Chilean Archipelago, while imbibed with towering forested slopes that fall into dark mysterious water, is also exposed to the full force of the westerlies and nearly devoid of community, thus making for a poor or impossible eclipse site. Once across the Andes however, and into Argentina, the weather improves significantly and the eclipse comes to its sunset ending near the resort town of El Calafate. The Andes block the flow of the westerlies, stripping them of their moisture and clouds, and leaving a drier and sunnier airflow to descend onto the plains of southern Argentina. No sunshine data are available for Argentina, but cloud-cover statistics (Table 3.8) show an encouraging average cloudiness for July at El Calafate of 55%. While this is about 10% higher than Tahiti, the data are similar to the values in the Cook Islands and parts of Polynesia.

The winter season brings cool temperatures, although nothing like the winters in the Northern Hemisphere. Average highs reach 6°C and average lows descend to a chilly (for the Southern Hemisphere) –5°C.

Because the Sun is close to setting during the eclipse, sight lines will have to be carefully arranged to avoid the distant mountains. That will be a tough challenge, as the eclipsed Sun is only 1° above the horizon, although the presence of several lakes aligned toward the west and northwest will help. The long view through the atmosphere will increase the probability that even a small amount of cloudiness will block totality.

3.7.7 Summary

French Polynesia is the clear-cut choice for the best weather prospects, but land-based sites are scarce and most observers will opt for a shipboard eclipse experience. Mangaia and Easter Island are the largest islands in the track, with more-or-less the same chances of sunshine—about 50%. If observing requirements dictate solid ground, then the choice is among one of these or the few reachable islands in the Tuamotus. Easter Island has, by far, the most developed infrastructure and the most convenient travel, but the small French Polynesian atolls offer the best weather. Easter Island, of course, has that aura of mystery that will more than compensate for the limited weather prospects.

Argentina is not a good choice if the eclipse alone is your goal. The very low altitude of the Sun, mountain-toothed horizon, and modest chances of sunshine suggest that more tropical destinations would be better.

TABLE 3.1

ELEMENTS OF THE TOTAL SOLAR ECLIPSE OF 2010 JULY 11

```
Equatorial Conjunction:      19:52:01.30 TDT    J.D. = 2455389.327793
  (Sun & Moon in R.A.)      (=19:50:55.11 UT)

Ecliptic Conjunction:        19:41:33.49 TDT    J.D. = 2455389.320527
  (Sun & Moon in Ec. Lo.)   (=19:40:27.31 UT)

    Instant of               19:34:37.63 TDT    J.D. = 2455389.315713
Greatest Eclipse:           (=19:33:31.45 UT)
```

Geocentric Coordinates of Sun & Moon at Greatest Eclipse (JPL DE200/LE200):

```
Sun:     R.A. = 07h23m57.621s      Moon:     R.A. = 07h23m15.844s
         Dec. =+22°02'10.95"                 Dec. =+21°22'29.30"
  Semi-Diameter =   15'43.94"         Semi-Diameter =   16'26.67"
  Eq.Hor.Par.  =      08.65"         Eq.Hor.Par.  =  1°00'20.87"
         Δ R.A. =    10.187s/h              Δ R.A. =   154.225s/h
         Δ Dec. =   -20.41"/h               Δ Dec. =   -515.89"/h
```

```
Lunar Radius    k1 = 0.2725076 (Penumbra)      Shift in       Δb =  0.00"
  Constants:    k2 = 0.2722810 (Umbra)       Lunar Position:  Δl =  0.00"

Geocentric Libration:   l =  -3.2°        Brown Lun. No. = 1083
(Optical + Physical)    b =   0.9°          Saros Series = 146 (27/76)
                        c =   6.6°                  nDot = -26.00 "/cy**2
```

Eclipse Magnitude = 1.05804   Gamma = -0.67877   ΔT =   66.2 s

Polynomial Besselian Elements for: 2010 Jul 11   20:00:00 TDT   (=t_0)

```
  n      x          y          d          l1         l2         μ

  0   0.0740999 -0.7170312 22.0357037  0.5344427 -0.0116561 118.614319
  1   0.5572523 -0.1366581 -0.0053410 -0.0000908 -0.0000904  15.000069
  2  -0.0000276 -0.0001121 -0.0000052 -0.0000124 -0.0000123   0.000002
  3  -0.0000090  0.0000024  0.0000000  0.0000000  0.0000000   0.000000
```

Tan f_1 = 0.0045988   Tan f_2 = 0.0045759

At time t_1 (decimal hours), each Besselian element is evaluated by:

$$a = a_0 + a_1 \ast t + a_2 \ast t^2 + a_3 \ast t^3 \quad (\text{or } a = \sum [a_n \ast t^n]; \, n = 0 \text{ to } 3)$$

where: a = x, y, d, l_1, l_2, or μ
         $t = t_1 - t_0$ (decimal hours) and t_0 = 20.00 TDT

The Besselian elements were derived from a least-squares fit to elements calculated at five uniformly spaced times over a 6-hour period centered at t_0. Thus, they are valid over the period 17.00 ≤ t_1 ≤ 23.00 TDT.

All times are expressed in Terrestrial Dynamical Time (TDT).

Saros Series 146: Member 27 of 76 eclipses in series.

TABLE 3.2

SHADOW CONTACTS AND CIRCUMSTANCES
TOTAL SOLAR ECLIPSE OF 2010 JULY 11

$$\Delta T = 66.2 \text{ s}$$
$$= 000°16'35.5"$$

		Terrestrial Dynamical Time h m s	Latitude	Ephemeris Longitude†	True Longitude*
External/Internal Contacts of Penumbra:	P_1	17:10:43.8	11°38.9'S	161°30.4'W	161°13.8'W
	P_4	21:58:20.5	36°47.3'S	075°48.4'W	075°31.9'W
Extreme North/South Limits of Penumbral Path:	N_1	18:00:45.9	04°40.3'N	179°17.7'E	179°34.3'E
	S_1	21:08:25.8	20°55.6'S	054°37.4'W	054°20.8'W
External/Internal Contacts of Umbra:	U_1	18:16:18.3	26°18.4'S	171°08.5'W	170°51.9'W
	U_2	18:19:36.0	27°25.4'S	171°23.1'W	171°06.5'W
	U_3	20:49:25.7	51°22.0'S	071°23.4'W	071°06.8'W
	U_4	20:52:47.2	50°21.7'S	071°03.1'W	070°46.5'W
Extreme North/South Limits of Umbral Path:	N_1	18:17:01.6	26°02.9'S	171°27.3'W	171°10.7'W
	S_1	18:18:54.5	27°40.6'S	171°04.8'W	170°48.2'W
	N_2	20:52:03.3	50°07.6'S	070°36.2'W	070°19.6'W
	S_2	20:50:08.0	51°35.5'S	071°50.3'W	071°33.7'W
Extreme Limits of Central Line:	C_1	18:17:56.7	26°51.4'S	171°16.1'W	170°59.5'W
	C_2	20:51:07.0	50°51.4'S	071°12.4'W	070°55.8'W
Instant of Greatest Eclipse:	G_0	19:34:37.6	19°44.9'S	122°09.1'W	121°52.5'W

Circumstances at Greatest Eclipse: Sun's Altitude = 47.1°   Path Width = 258.6 km
                                    Sun's Azimuth = 13.5°   Central Duration = 05m20.2s

---

† Ephemeris Longitude is the terrestrial dynamical longitude assuming a uniformly rotating Earth.
* True Longitude is calculated by correcting the Ephemeris Longitude for the non-uniform rotation of Earth.
  (T.L. = E.L. + 1.002738*ΔT/240, where ΔT(in seconds) = TDT - UT)

Note: Longitude is measured positive to the East.

Because ΔT is not known in advance, the value used in the predictions is an extrapolation based on pre-2009 measurements. The actual value is expected to fall within ±0.3 seconds of the estimated ΔT used here.

TABLE 3.3

PATH OF THE UMBRAL SHADOW
TOTAL SOLAR ECLIPSE OF 2010 JULY 11

$\Delta T = 66.2$ s

Universal Time	Northern Limit Latitude	Northern Limit Longitude	Southern Limit Latitude	Southern Limit Longitude	Central Line Latitude	Central Line Longitude	Sun Alt °	Path Width km	Central Durat.
Limits	26°02.9'S	171°10.7'W	27°40.6'S	170°48.2'W	26°51.4'S	170°59.5'W	0	179	02m42.4s
18:20	20°50.8'S	157°59.3'W	23°57.1'S	161°17.4'W	22°19.8'S	159°27.6'W	12	197	03m15.1s
18:25	18°54.4'S	152°08.8'W	21°29.5'S	154°04.9'W	20°10.7'S	153°03.5'W	19	209	03m37.8s
18:30	17°44.8'S	147°59.6'W	20°10.9'S	149°29.3'W	18°57.0'S	148°42.5'W	24	219	03m55.0s
18:35	16°58.8'S	144°38.3'W	19°20.7'S	145°54.6'W	18°09.2'S	145°15.0'W	28	227	04m09.5s
18:40	16°28.2'S	141°46.0'W	18°47.8'S	142°54.2'W	17°37.6'S	142°19.0'W	32	235	04m22.1s
18:45	16°09.0'S	139°13.5'W	18°27.2'S	140°16.4'W	17°17.7'S	139°44.0'W	35	241	04m33.3s
18:50	15°58.9'S	136°55.3'W	18°16.1'S	137°54.5'W	17°07.1'S	137°24.2'W	37	247	04m43.1s
18:55	15°56.2'S	134°48.0'W	18°12.7'S	135°44.6'W	17°04.1'S	135°15.7'W	39	252	04m51.7s
19:00	16°00.1'S	132°49.1'W	18°16.0'S	133°43.8'W	17°07.7'S	133°15.9'W	41	256	04m59.1s
19:05	16°09.8'S	130°56.9'W	18°25.2'S	131°50.1'W	17°17.2'S	131°23.1'W	43	259	05m05.4s
19:10	16°24.8'S	129°10.0'W	18°39.8'S	130°02.0'W	17°31.9'S	129°35.6'W	44	261	05m10.5s
19:15	16°44.7'S	127°27.2'W	18°59.3'S	128°18.3'W	17°51.7'S	127°52.5'W	45	262	05m14.6s
19:20	17°09.4'S	125°47.7'W	19°23.6'S	126°38.0'W	18°16.1'S	126°12.6'W	46	262	05m17.5s
19:25	17°38.6'S	124°10.6'W	19°52.4'S	125°00.1'W	18°45.1'S	124°35.1'W	47	261	05m19.4s
19:30	18°12.2'S	122°35.1'W	20°25.7'S	123°23.8'W	19°18.6'S	122°59.4'W	47	260	05m20.3s
19:35	18°50.3'S	121°00.7'W	21°03.5'S	121°48.5'W	19°56.5'S	121°24.5'W	47	258	05m20.0s
19:40	19°32.9'S	119°26.5'W	21°45.8'S	120°13.4'W	20°38.9'S	119°49.9'W	47	255	05m18.8s
19:45	20°20.0'S	117°51.9'W	22°32.7'S	118°37.7'W	21°25.9'S	118°14.8'W	46	252	05m16.5s
19:50	21°11.9'S	116°16.2'W	23°24.5'S	117°00.6'W	22°17.7'S	116°38.5'W	46	249	05m13.3s
19:55	22°08.7'S	114°38.5'W	24°21.4'S	115°21.3'W	23°14.5'S	115°00.0'W	45	245	05m09.1s
20:00	23°10.9'S	112°57.8'W	25°23.8'S	113°38.6'W	24°16.8'S	113°18.4'W	43	241	05m03.9s
20:05	24°18.9'S	111°13.1'W	26°32.3'S	111°51.5'W	25°25.0'S	111°32.6'W	42	237	04m57.8s
20:10	25°33.4'S	109°23.0'W	27°47.7'S	109°58.3'W	26°39.9'S	109°41.0'W	40	233	04m50.7s
20:15	26°55.2'S	107°25.6'W	29°10.9'S	107°57.0'W	28°02.3'S	107°41.8'W	38	228	04m42.7s
20:20	28°25.7'S	105°18.6'W	30°43.3'S	105°44.8'W	29°33.7'S	105°32.3'W	36	224	04m33.7s
20:25	30°06.6'S	102°58.5'W	32°27.2'S	103°17.8'W	31°16.0'S	103°09.0'W	33	219	04m23.6s
20:30	32°00.8'S	100°20.1'W	34°25.8'S	100°29.7'W	33°12.2'S	100°26.0'W	30	214	04m12.3s
20:35	34°12.7'S	097°14.8'W	36°45.0'S	097°09.4'W	35°27.5'S	097°13.7'W	26	209	03m59.4s
20:40	36°51.0'S	093°26.0'W	39°36.2'S	092°54.4'W	38°11.7'S	093°12.9'W	21	204	03m44.4s
20:45	40°16.5'S	088°13.0'W	43°32.5'S	086°39.3'W	41°50.5'S	087°33.0'W	15	197	03m25.7s
20:50	46°09.3'S	078°12.1'W	–	–	50°24.0'S	071°53.8'W	1	183	02m47.2s
Limits	50°07.6'S	070°19.6'W	51°35.5'S	071°33.7'W	50°51.4'S	070°55.8'W	0	183	02m45.4s

TABLE 3.4

PHYSICAL EPHEMERIS OF THE UMBRAL SHADOW
TOTAL SOLAR ECLIPSE OF 2010 JULY 11

$\Delta T = 66.2$ s

Universal Time	Central Line Latitude	Central Line Longitude	Diameter Ratio	Eclipse Obscur.	Sun Alt °	Sun Azm °	Path Width km	Major Axis km	Minor Axis km	Umbra Veloc. km/s	Central Durat.
18:16.8	26°51.4'S	170°59.5'W	1.0441	1.0902	0.0	65.1	179.2	–	147.2	–	02m42.4s
18:20	22°19.8'S	159°27.6'W	1.0480	1.0982	12.0	59.8	197.0	761.8	159.4	3.160	03m15.1s
18:25	20°10.7'S	153°03.5'W	1.0502	1.1029	19.2	56.5	209.5	504.6	166.5	1.842	03m37.8s
18:30	18°57.0'S	148°42.5'W	1.0517	1.1062	24.2	53.9	219.2	416.3	171.4	1.385	03m55.0s
18:35	18°09.2'S	145°15.0'W	1.0529	1.1087	28.3	51.3	227.5	368.9	175.1	1.138	04m09.5s
18:40	17°37.6'S	142°19.0'W	1.0539	1.1107	31.7	48.8	234.8	338.7	178.1	0.980	04m22.1s
18:45	17°17.7'S	139°44.0'W	1.0547	1.1124	34.6	46.2	241.3	317.8	180.7	0.871	04m33.3s
18:50	17°07.1'S	137°24.2'W	1.0554	1.1139	37.1	43.5	247.0	302.5	182.9	0.791	04m43.1s
18:55	17°04.1'S	135°15.7'W	1.0560	1.1151	39.3	40.6	251.8	291.0	184.7	0.732	04m51.7s
19:00	17°07.7'S	133°15.9'W	1.0565	1.1162	41.2	37.6	255.7	282.1	186.2	0.687	04m59.1s
19:05	17°17.2'S	131°23.1'W	1.0569	1.1171	42.9	34.4	258.6	275.3	187.6	0.654	05m05.4s
19:10	17°31.9'S	129°35.6'W	1.0573	1.1178	44.2	31.0	260.7	270.0	188.6	0.629	05m10.5s
19:15	17°51.7'S	127°52.5'W	1.0575	1.1184	45.3	27.5	261.8	266.1	189.5	0.612	05m14.6s
19:20	18°16.1'S	126°12.6'W	1.0578	1.1188	46.2	23.9	262.0	263.2	190.2	0.602	05m17.5s
19:25	18°45.1'S	124°35.1'W	1.0579	1.1192	46.7	20.1	261.4	261.4	190.7	0.597	05m19.4s
19:30	19°18.6'S	122°59.4'W	1.0580	1.1194	47.1	16.3	260.0	260.5	191.0	0.597	05m20.3s
19:35	19°56.5'S	121°24.5'W	1.0580	1.1195	47.1	12.4	257.9	260.4	191.1	0.603	05m20.0s
19:40	20°38.9'S	119°49.9'W	1.0580	1.1194	46.9	8.4	255.3	261.3	191.0	0.613	05m18.8s
19:45	21°25.9'S	118°14.8'W	1.0579	1.1193	46.4	4.6	252.3	263.0	190.8	0.628	05m16.5s
19:50	22°17.7'S	116°38.5'W	1.0578	1.1190	45.7	0.7	248.8	265.6	190.3	0.649	05m13.3s
19:55	23°14.5'S	115°00.0'W	1.0576	1.1185	44.7	356.9	245.0	269.4	189.7	0.675	05m09.1s
20:00	24°16.8'S	113°18.4'W	1.0573	1.1180	43.4	353.2	241.1	274.4	188.9	0.707	05m03.9s
20:05	25°25.0'S	111°32.6'W	1.0570	1.1173	41.9	349.6	236.9	280.8	187.9	0.748	04m57.8s
20:10	26°39.9'S	109°41.0'W	1.0566	1.1164	40.1	346.1	232.7	289.2	186.6	0.798	04m50.7s
20:15	28°02.3'S	107°41.8'W	1.0561	1.1153	38.0	342.7	228.3	300.0	185.0	0.860	04m42.7s
20:20	29°33.7'S	105°32.3'W	1.0555	1.1141	35.5	339.3	223.8	314.1	183.2	0.939	04m33.7s
20:25	31°16.0'S	103°09.0'W	1.0548	1.1126	32.8	335.9	219.1	333.1	180.9	1.043	04m23.6s
20:30	33°12.2'S	100°26.0'W	1.0539	1.1108	29.6	332.5	214.3	359.8	178.2	1.186	04m12.3s
20:35	35°27.5'S	097°13.7'W	1.0529	1.1086	25.8	328.9	209.2	400.1	175.0	1.397	03m59.4s
20:40	38°11.7'S	093°12.9'W	1.0516	1.1058	21.2	324.8	203.6	469.3	170.8	1.754	03m44.4s
20:45	41°50.5'S	087°33.0'W	1.0497	1.1019	15.1	319.7	196.9	628.5	165.0	2.561	03m25.7s
20:50	50°24.0'S	071°53.8'W	1.0452	1.0925	0.8	307.2	183.5	9822.8	150.7	48.256	02m47.2s
20:50.0	50°51.4'S	070°55.8'W	1.0450	1.0920	0.0	306.5	182.8	–	149.9	–	02m45.4s

TABLE 3.5

LOCAL CIRCUMSTANCES ON THE CENTRAL LINE
TOTAL SOLAR ECLIPSE OF 2010 JULY 11

$\Delta T = 66.2$ s

Central Line Maximum Eclipse			First Contact				Second Contact			Third Contact			Fourth Contact			
U.T.	Durat.	Alt	U.T.	P	V	Alt	U.T.	P	V	U.T.	P	V	U.T.	P	V	Alt
18:20	03m15.1s	12	–	–	–	–	18:18:23	97	217	18:21:38	277	37	19:34:32	98	229	26
18:25	03m37.8s	19	17:15:54	276	30	5	18:23:12	98	220	18:26:49	278	40	19:44:45	100	235	34
18:30	03m55.0s	24	17:17:46	277	32	10	18:28:03	99	223	18:31:58	279	43	19:53:30	101	242	39
18:35	04m09.5s	28	17:20:04	278	34	14	18:32:56	100	226	18:37:05	280	47	20:01:26	103	248	42
18:40	04m22.1s	32	17:22:39	279	36	17	18:37:49	101	230	18:42:12	281	50	20:08:48	104	254	45
18:45	04m33.3s	35	17:25:26	279	38	20	18:42:44	102	233	18:47:17	282	54	20:15:42	105	261	47
18:50	04m43.1s	37	17:28:25	280	40	22	18:47:39	103	237	18:52:22	283	58	20:22:11	106	267	49
18:55	04m51.7s	39	17:31:33	281	43	24	18:52:35	104	241	18:57:26	284	62	20:28:18	108	273	50
19:00	04m59.1s	41	17:34:51	282	45	27	18:57:31	105	245	19:02:30	285	67	20:34:06	109	279	50
19:05	05m05.4s	43	17:38:18	282	48	29	19:02:28	106	249	19:07:33	286	71	20:39:36	110	285	51
19:10	05m10.5s	44	17:41:54	283	51	30	19:07:25	107	254	19:12:36	287	76	20:44:51	111	291	50
19:15	05m14.6s	45	17:45:40	284	54	32	19:12:23	108	258	19:17:38	288	80	20:49:51	111	296	50
19:20	05m17.5s	46	17:49:35	285	57	34	19:17:22	109	263	19:22:39	289	85	20:54:37	112	301	49
19:25	05m19.4s	47	17:53:40	286	61	35	19:22:21	109	268	19:27:40	290	90	20:59:12	113	306	48
19:30	05m20.3s	47	17:57:56	287	65	36	19:27:20	110	273	19:32:40	291	95	21:03:36	114	310	47
19:35	05m20.0s	47	18:02:24	288	68	38	19:32:20	111	278	19:37:40	291	100	21:07:50	114	314	46
19:40	05m18.8s	47	18:07:02	288	72	39	19:37:21	112	282	19:42:39	292	104	21:11:55	115	317	44
19:45	05m16.5s	46	18:11:53	289	76	39	19:42:22	113	287	19:47:38	293	109	21:15:52	115	320	43
19:50	05m13.3s	46	18:16:56	290	81	40	19:47:23	113	292	19:52:37	293	114	21:19:41	116	323	41
19:55	05m09.1s	45	18:22:13	291	85	40	19:52:25	114	296	19:57:34	294	118	21:23:22	116	325	39
20:00	05m03.9s	43	18:27:44	292	90	41	19:57:28	114	300	20:02:32	294	122	21:26:57	116	327	37
20:05	04m57.8s	42	18:33:28	292	95	41	20:02:31	115	304	20:07:29	295	126	21:30:25	116	329	35
20:10	04m50.7s	40	18:39:28	293	100	40	20:07:34	115	308	20:12:25	295	129	21:33:46	116	331	32
20:15	04m42.7s	38	18:45:44	293	105	39	20:12:38	115	311	20:17:21	295	133	21:37:00	116	332	30
20:20	04m33.7s	36	18:52:16	294	110	38	20:17:43	116	314	20:22:17	296	136	21:40:07	116	333	27
20:25	04m23.6s	33	18:59:06	294	115	37	20:22:48	116	317	20:27:12	296	138	21:43:05	116	334	24
20:30	04m12.3s	30	19:06:17	295	119	35	20:27:54	116	320	20:32:06	296	141	21:45:54	116	335	20
20:35	03m59.4s	26	19:13:52	295	124	32	20:33:00	115	322	20:36:59	295	143	21:48:29	116	335	16
20:40	03m44.4s	21	19:21:59	295	129	28	20:38:08	115	324	20:41:52	295	145	21:50:45	115	335	12
20:45	03m25.7s	15	19:31:00	294	133	22	20:43:17	114	325	20:46:43	294	146	21:52:24	114	334	6
20:50	02m47.2s	1	19:44:30	292	138	8	20:48:36	112	325	20:51:23	292	146	–	–	–	–
20:50	02m45.5s	1	19:44:55	292	138	8	20:48:38	112	325	20:51:23	292	145	–	–	–	–

TABLE 3.6

TOPOCENTRIC DATA AND PATH CORRECTIONS DUE TO LUNAR LIMB PROFILE
TOTAL SOLAR ECLIPSE OF 2010 JULY 11

ΔT = 66.2 s

Universal Time	Moon Topo H.P. "	Moon Topo S.D. "	Moon Rel. Ang.V "/s	Topo Lib. Long °	Sun Alt. °	Sun Az. °	Path Az. °	North Limit P.A. °	North Limit Int. '	North Limit Ext. '	South Limit Int. '	South Limit Ext. '	Central Durat. Corr. s
18:20	3632.8	989.2	0.464	-2.53	12.0	59.8	68.6	6.6	-0.3	0.5	0.6	-2.3	-2.2
18:25	3640.7	991.3	0.435	-2.58	19.2	56.5	71.9	7.7	-0.2	0.6	0.3	-2.4	-2.7
18:30	3646.0	992.8	0.416	-2.62	24.2	53.9	74.9	8.7	-0.1	0.7	0.2	-3.0	-3.3
18:35	3650.2	993.9	0.401	-2.66	28.3	51.3	77.9	9.8	0.1	0.7	0.1	-3.1	-3.9
18:40	3653.6	994.8	0.388	-2.71	31.7	48.8	80.9	10.8	0.2	0.7	-0.1	-3.0	-4.1
18:45	3656.4	995.6	0.378	-2.75	34.6	46.2	83.9	11.8	0.3	1.0	-0.3	-2.5	-4.6
18:50	3658.8	996.3	0.370	-2.79	37.1	43.5	87.0	12.8	0.3	1.1	-0.3	-3.0	-4.9
18:55	3660.9	996.8	0.362	-2.83	39.3	40.6	90.2	13.8	0.3	1.5	-0.3	-3.3	-5.2
19:00	3662.6	997.3	0.357	-2.88	41.2	37.6	93.4	14.8	0.2	1.4	-0.4	-3.2	-5.4
19:05	3664.1	997.7	0.352	-2.92	42.9	34.4	96.6	15.8	0.2	1.1	-0.4	-3.3	-5.5
19:10	3665.3	998.0	0.348	-2.96	44.2	31.0	99.8	16.8	0.2	0.9	-0.2	-3.8	-5.6
19:15	3666.3	998.3	0.345	-3.00	45.3	27.5	102.9	17.8	0.2	0.7	-0.2	-4.2	-5.5
19:20	3667.1	998.5	0.343	-3.05	46.2	23.9	106.0	18.7	0.2	0.9	-0.1	-4.3	-5.8
19:25	3667.6	998.6	0.342	-3.09	46.7	20.1	108.9	19.6	0.2	1.2	-0.1	-4.0	-6.0
19:30	3668.0	998.7	0.342	-3.13	47.1	16.3	111.7	20.4	0.3	1.4	0.0	-3.3	-6.2
19:35	3668.1	998.8	0.342	-3.18	47.1	12.4	114.3	21.2	0.3	1.3	0.0	-2.7	-6.3
19:40	3668.0	998.7	0.344	-3.22	46.9	8.4	116.8	22.0	0.4	1.2	0.0	-2.2	-6.3
19:45	3667.7	998.7	0.346	-3.26	46.4	4.6	119.0	22.7	0.5	1.6	0.0	-2.6	-6.3
19:50	3667.2	998.5	0.348	-3.30	45.7	0.7	121.1	23.3	0.6	1.7	-0.0	-2.8	-6.4
19:55	3666.5	998.4	0.352	-3.35	44.7	356.9	123.0	23.9	0.7	1.7	-0.0	-2.8	-6.3
20:00	3665.6	998.1	0.356	-3.39	43.4	353.2	124.6	24.4	0.7	1.5	-0.1	-2.7	-6.1
20:05	3664.4	997.8	0.361	-3.43	41.9	349.6	126.1	24.8	0.8	1.3	-0.1	-2.5	-5.9
20:10	3663.0	997.4	0.367	-3.48	40.1	346.1	127.4	25.2	0.9	1.2	-0.1	-2.3	-5.8
20:15	3661.2	996.9	0.375	-3.52	38.0	342.7	128.4	25.5	1.0	1.4	-0.1	-2.2	-5.6
20:20	3659.1	996.4	0.383	-3.56	35.5	339.3	129.3	25.6	1.0	1.5	-0.1	-2.2	-5.5
20:25	3656.6	995.7	0.392	-3.61	32.8	335.9	129.9	25.7	1.0	1.5	-0.1	-2.3	-5.4
20:30	3653.6	994.9	0.404	-3.65	29.6	332.5	130.3	25.6	1.0	1.5	-0.1	-2.2	-5.2
20:35	3649.9	993.9	0.417	-3.69	25.8	328.9	130.5	25.4	0.9	1.4	-0.1	-2.2	-5.1
20:40	3645.3	992.6	0.434	-3.73	21.2	324.8	130.3	25.0	0.8	1.2	-0.1	-2.5	-5.1
20:45	3638.9	990.9	0.456	-3.78	15.1	319.7	129.7	24.3	0.7	1.6	-0.0	-2.8	-5.0
20:50	3623.6	986.6	0.511	-3.82	0.8	307.2	126.9	22.0	0.7	1.6	-0.0	-2.8	-5.0

TABLE 3.7
LOCAL CIRCUMSTANCES FOR PACIFIC OCEAN & SOUTH AMERICA
TOTAL SOLAR ECLIPSE OF 2010 JULY 11

Location Name	Latitude	Longitude	Elev. m	First Contact U.T. h m s	P °	V °	Alt °	Second Contact U.T. h m s	P °	V °	Alt °	Third Contact U.T. h m s	P °	V °	Alt °	Fourth Contact U.T. h m s	P °	V °	Alt °	Maximum Eclipse U.T. h m s	P °	V °	Alt °	Azm °	Eclip. Mag.	Eclip. Obs.	Umbral Depth	Umbral Durat.
PACIFIC OCEAN																												
COOK ISLANDS																												
Mangaia Is.	21°55'S	157°55'W	--	17:15:06.6	276	30	1	18:19:27.3	105	226		18:22:45.4	269	30		19:36:50.8	98	230	28	18:21:06.0	7	128	14	59	1.048	1.000	0.856	03m18s
Mauke Is.	20°09'S	157°23'W	--	17:13:48.5	274	26	1									19:36:03.4	101	231	29	18:19:57.8	187	306	15	59	0.988	0.991		
Rarotonga Is.	21°14'S	159°46'W	--													19:32:53.5	100	229	26	18:18:39.5	186	305	12	60	0.993	0.996		
FIJI																												
Lautoka	17°37'S	177°27'E	--													19:01:53.4	116	227	4	18:43 Rise	--	--	0	67	0.286	0.178		
Suva	18°08'S	178°25'E	6													19:03:19.0	115	226	5	18:40 Rise	--	--	0	67	0.351	0.239		
FRENCH POLYNESIA																												
Bora-Bora	16°30'S	151°45'W	--	17:13:51.9	272	23	7									19:43:56.2	105	238	37	18:23:24.0	188	307	22	57	0.941	0.937		
Moorea Is.	17°32'S	149°40'W	--	17:15:53.4	274	28	9									19:44:49.8	103	241	39	18:27:24.0	188	310	24	55	0.983	0.987		
Papeete, Tahiti	17°32'S	149°34'W	2	17:15:57.6	274	28	9									19:50:04.1	103	241	39	18:27:22.9	188	311	24	55	0.984	0.987		
Rikitea	23°08'S	134°57'W	--	17:39:16.0	289	58	23									20:33:32.3	100	270	44	19:02:09.3	14	159	36	36	0.887	0.869		
Hikueru	17°35'S	142°39'W	2	17:22:14.1	278	35	16	18:37:08.7	95	224		18:41:28.4	286	35	55	20:07:50.6	104	253	45	18:39:18.0	191	319	31	49	1.054	1.000	0.910	04m20s
Tatakoto	17°22'S	138°26'W	2	17:27:13.0	280	40	21	18:45:37.7	112	245		18:50:12.4	273	40	47	20:19:26.6	106	264	48	18:47:54.6	12	146	36	45	1.055	1.000	0.835	04m35s
SAMOA, WESTERN																												
Apia	13°50'S	171°44'W	--													19:06:45.0	116	228	16	18:05:50.7	185	290	3	67	0.664	0.588		
TONGA																												
Nuku'alofa	21°08'S	175°12'W	--													19:12:35.6	106	224	11	18:20 Rise	--	--	0	67	0.768	0.716		
EASTER ISLAND (ISLA DE PASCUA, CHILE)																												
Hanga Roa	27°09'S	109°26'W	--	18:40:35.9	293	101	40	20:08:30.0	129	322		20:13:10.9	282	116		21:34:16.3	116	330	32	20:10:50.6	25	219	40	346	1.056	1.000	0.767	04m41s
GALAPAGOS ISLANDS (ECUADOR)																												
Isla Santa Cruz	00°38'S	090°23'W	--	20:07:57.9	221	97	53									20:39:40.7	198	82	46	20:23:55.4	209	90	50	306	0.020	0.003		
SOUTH AMERICA																												
ARGENTINA																												
Buenos Aires	34°36'S	058°27'W	27	20:13:43.8	260	128	7									--				20:57 Set	--	--	0	297	0.446	0.336		
Córdoba	31°24'S	064°11'W	--	20:11:52.8	260	128	13									21:04:11.5	204	78	4	20:49:53.8	22	235	1	308	1.045	1.000	0.902	02m47s
El Calafate	50°20'S	072°15'W	--	19:44:15.1	292	138	9	20:48:30.1	118	331		20:51:17.1	286	140		--				20:55 Set	--	--	0	297	0.437	0.327		
La Plata	34°55'S	057°57'W	--	20:13:44.6	260	129	7									--				20:57 Set	--	--	0	297	0.448	0.338		
Lomas de Zamora	34°46'S	058°24'W	--	20:13:31.5	261	129	7									--				20:57 Set	--	--	0	298	0.452	0.343		
Mar del Plata	38°00'S	057°33'W	--	20:09:36.2	266	131	6									--				20:45 Set	--	--	0	298	0.452	0.343		
Mendoza	32°53'S	068°49'W	800	20:03:39.3	267	130	17									--				21:01:56.8	204	75	7	302	0.542	0.444		
Rosario	32°57'S	060°40'W	--	20:13:52.7	259	128	10									--				21:04:48.9	204	77	1	297	0.438	0.329		
San Miguel de T*	26°49'S	065°13'W	--	20:17:46.1	252	124	16									--				21:04:26.8	205	82	7	299	0.319	0.209		
BOLIVIA																												
La Paz	16°30'S	068°09'W	3658	20:39:13.3	226	108	20									21:22:52.5	186	73	10	21:01:23.3	206	91	15	299	0.061	0.018		
BRAZIL																												
Jardim	21°28'S	056°09'W	--	20:50:53.5	220	105	5									--				21:06:58.9	204	90	2	294	0.039	0.009		
Porto Alegre	30°04'S	051°11'W	10	20:29:31.6	243	120	2									--				20:39 Set	--	--	0	295	0.096	0.036		
CHILE																												
Concepción	36°50'S	073°03'W	--	19:53:54.7	277	133	18									--				20:58:25.7	204	69	8	306	0.707	0.641		
Santiago	33°27'S	070°40'W	--	20:00:24.7	270	131	18									--				21:00:51.8	205	74	8	303	0.583	0.492		
Valparaíso	33°02'S	071°38'W	41	19:59:27.0	270	130	19									--				21:00:27.3	205	74	9	304	0.585	0.494		
PARAGUAY																												
Asunción	25°16'S	057°40'W	139	20:33:13.2	238	117	8									--				21:06:47.6	204	86	1	295	0.173	0.085		
PERU																												
Lima	12°03'S	077°03'W	120	20:24:15.3	231	110	32									21:18:56.8	184	71	21	20:52:11.9	208	91	27	302	0.081	0.028		
URUGUAY																												
Montevideo	34°53'S	056°11'W	22	20:15:29.2	259	128	5									--				20:48 Set	--	--	0	297	0.368	0.256		

Table 3.8: Climate Statistics for July along the Total Eclipse Path

July Climate Statistics	Percent of possible sunshine	Percent Frequency of Sky Condition					Calculated Cloudiness	July Precip.	Ave. no. days with rain	Ave. no. days with tstms	Ave. no. days with snow	Days with fog	Ave. July High	Ave. July Low
		Clear	Few	Scattered	Broken	Overcast	Obscured	%	mm					(°C)	(°C)
Cook Islands															
Mangaia Island*		0.3	16.7	21.7	45.3	16.0	0.0	63	159	15.3	0.4			25	19
Rarotonga	52	0.0	13.1	25.9	50.4	10.6	0.0	63	104	16.5	0.3		0.1	25	20
Mauke		2.2	30.3	15.1	42.1	10.3	0.0	55	115	11.9	0.2			26	21
Tahiti															
Papeete	68	0.2	38.1	25.1	32.9	3.7	0.0	46	53	7.4	0.4			29	21
Tuamotu Atoll															
Takaroa	66	0.2	38.7	30.2	26.7	4.1	0.2	44							
Rikitea	45	0.0	26.0	28.0	40.1	5.9	0.0	54							
Hao*		0.4	20.2	36.1	41.8	1.5	0.0	53	76.1	10.2				27	23
Hereheretue	68	0.4	20.2	36.1	41.8	1.5	0.0	53	72.3	12				27	22
Austral Islands															
Rapa Iti	38	0.0	8.1	13.9	57.9	20.1	0.0	73	207					21	16
Tubuai	51	0.3	19.6	16.8	52.4	10.8	0.0	62	144.9					24	18
Chile															
Easter Island *	50	0.0	7.4	20.3	57.1	15.2	0.0	70	111	18.1				21	16
Puerto Natales		0.0	13.9	25.0	41.7	16.7	2.8	65	33	11.4				4	-1
Argentina															
El Calafate*		7.8	18.8	24.7	34.4	14.3	0.0	55	42	3.6	0	3.4	1	6	-5

* This location is in the path of totality.

FIGURE 3.1: ORTHOGRAPHIC PROJECTION MAP OF THE ECLIPSE PATH
Total Solar Eclipse of 2010 Jul 11

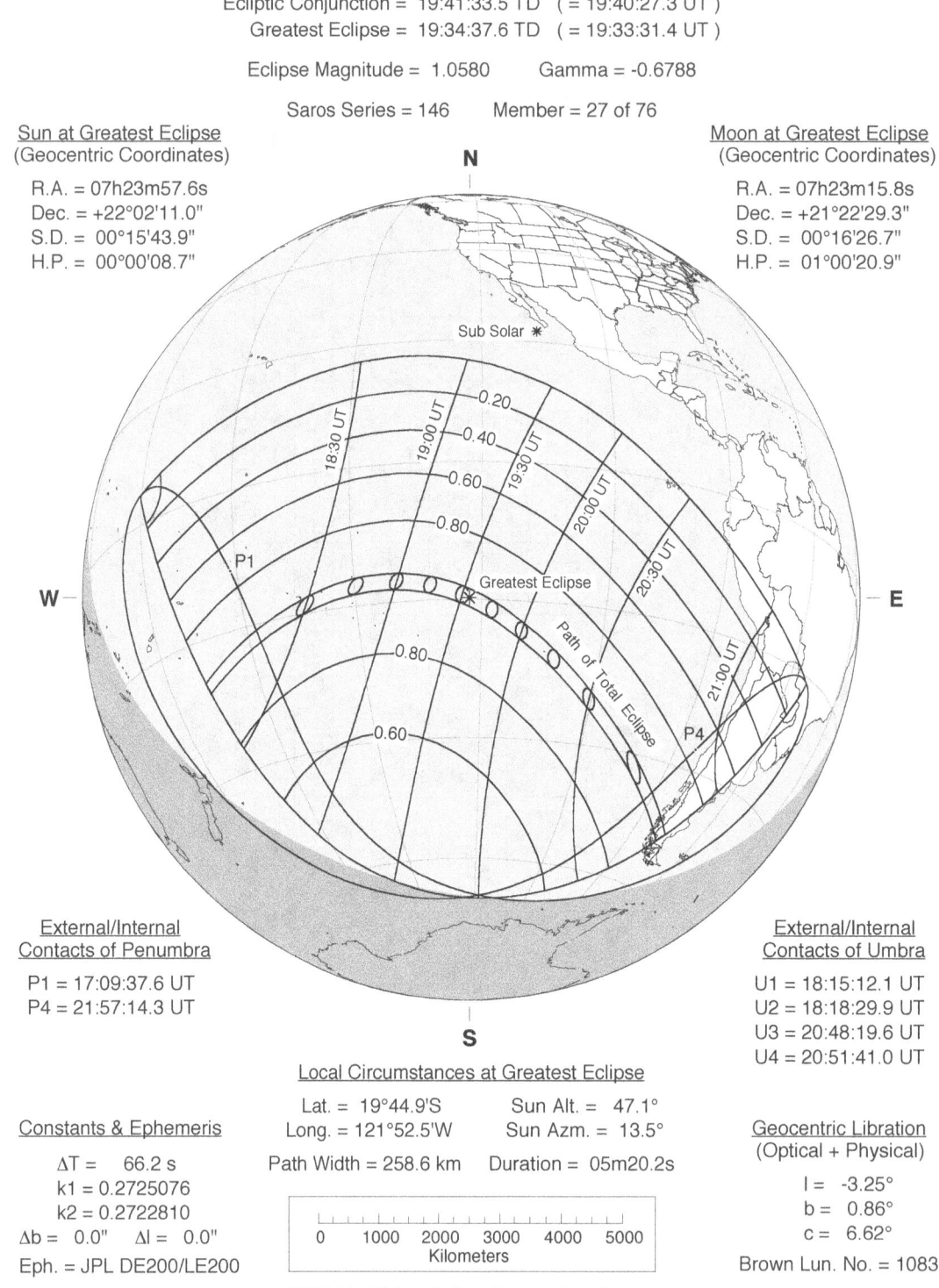

NASA 2010 Eclipse Bulletin, Espenak & Anderson

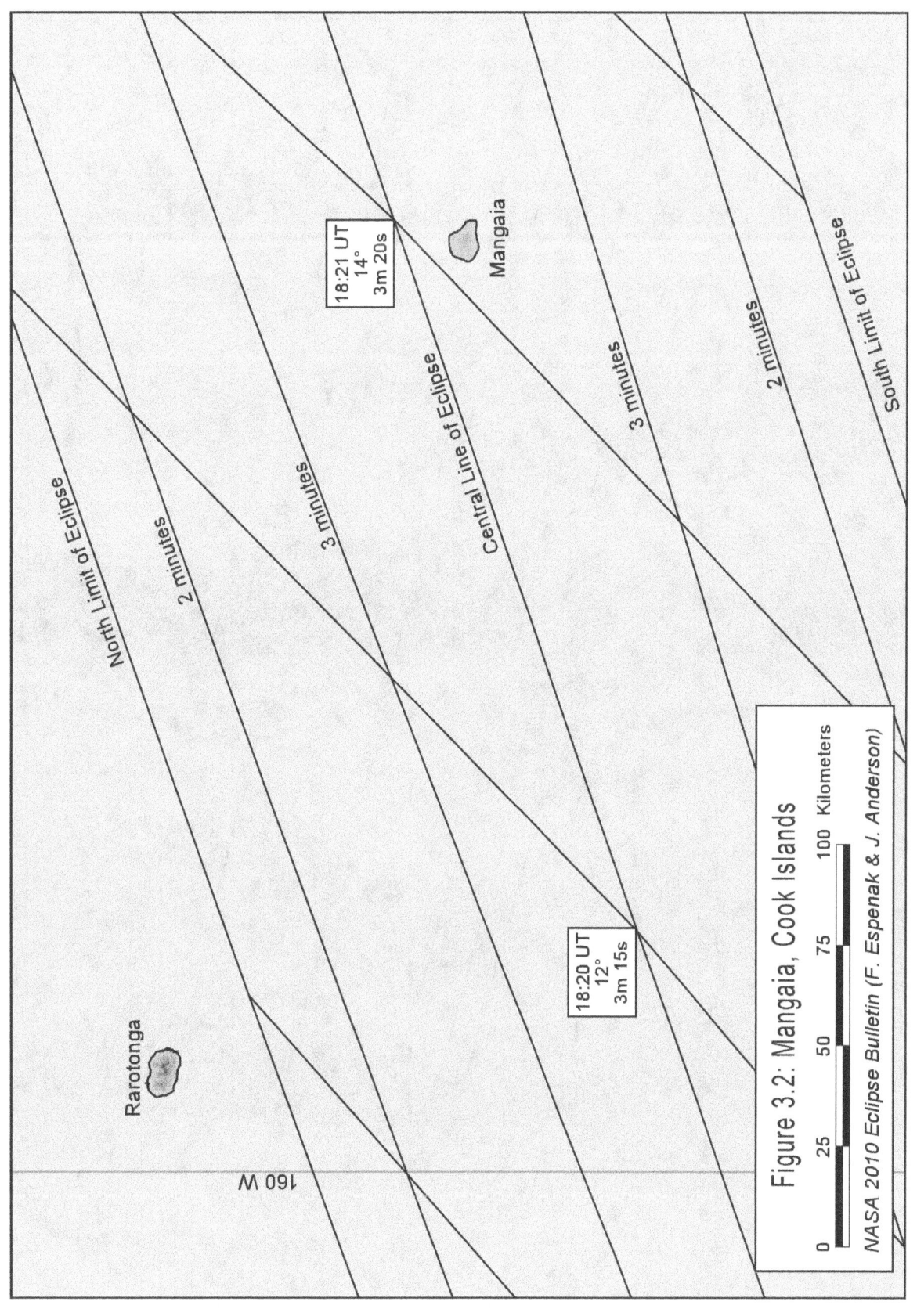

Figure 3.2: Mangaia, Cook Islands

Total Solar Eclipse of 2010 July 11

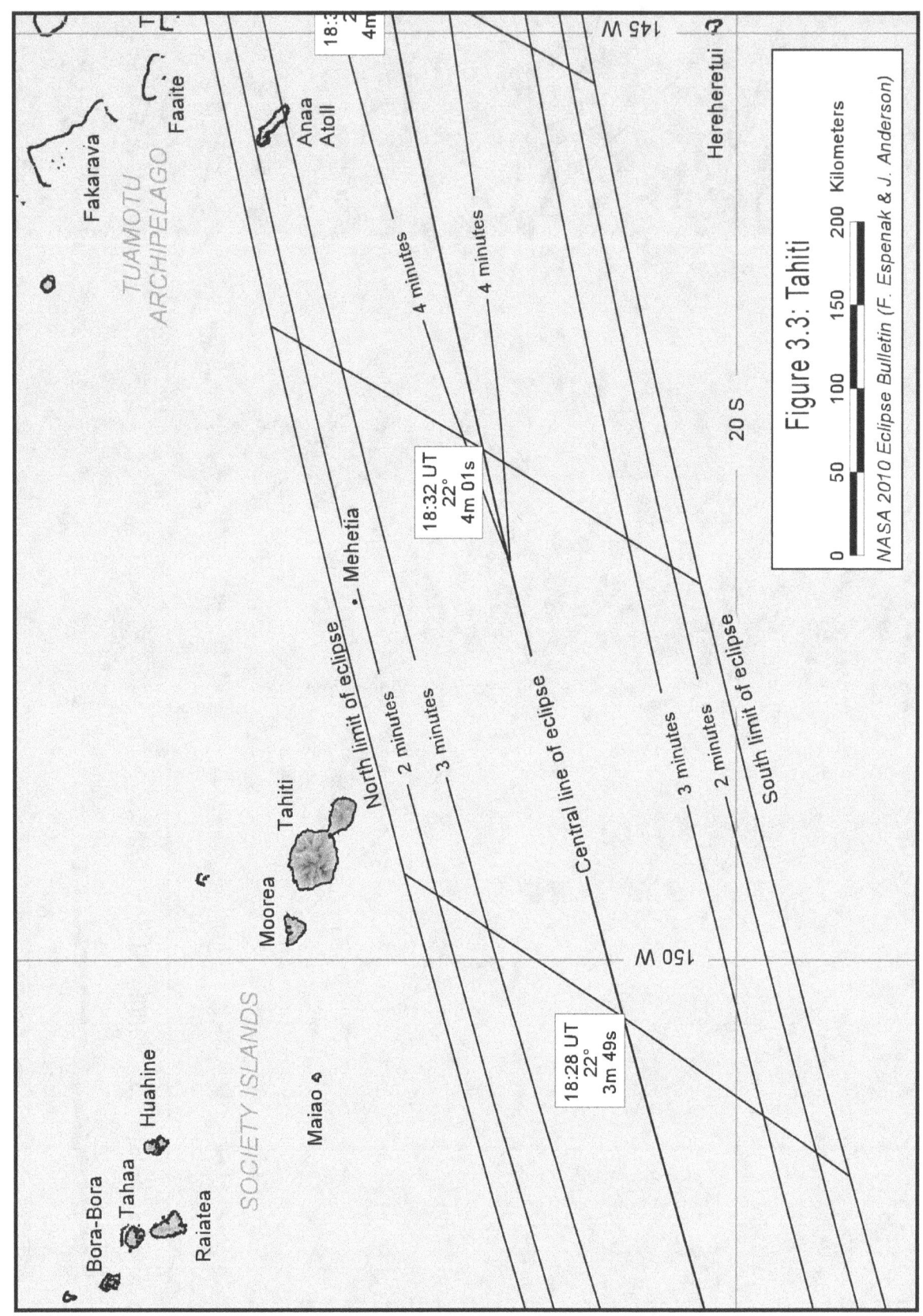

Figure 3.3: Tahiti

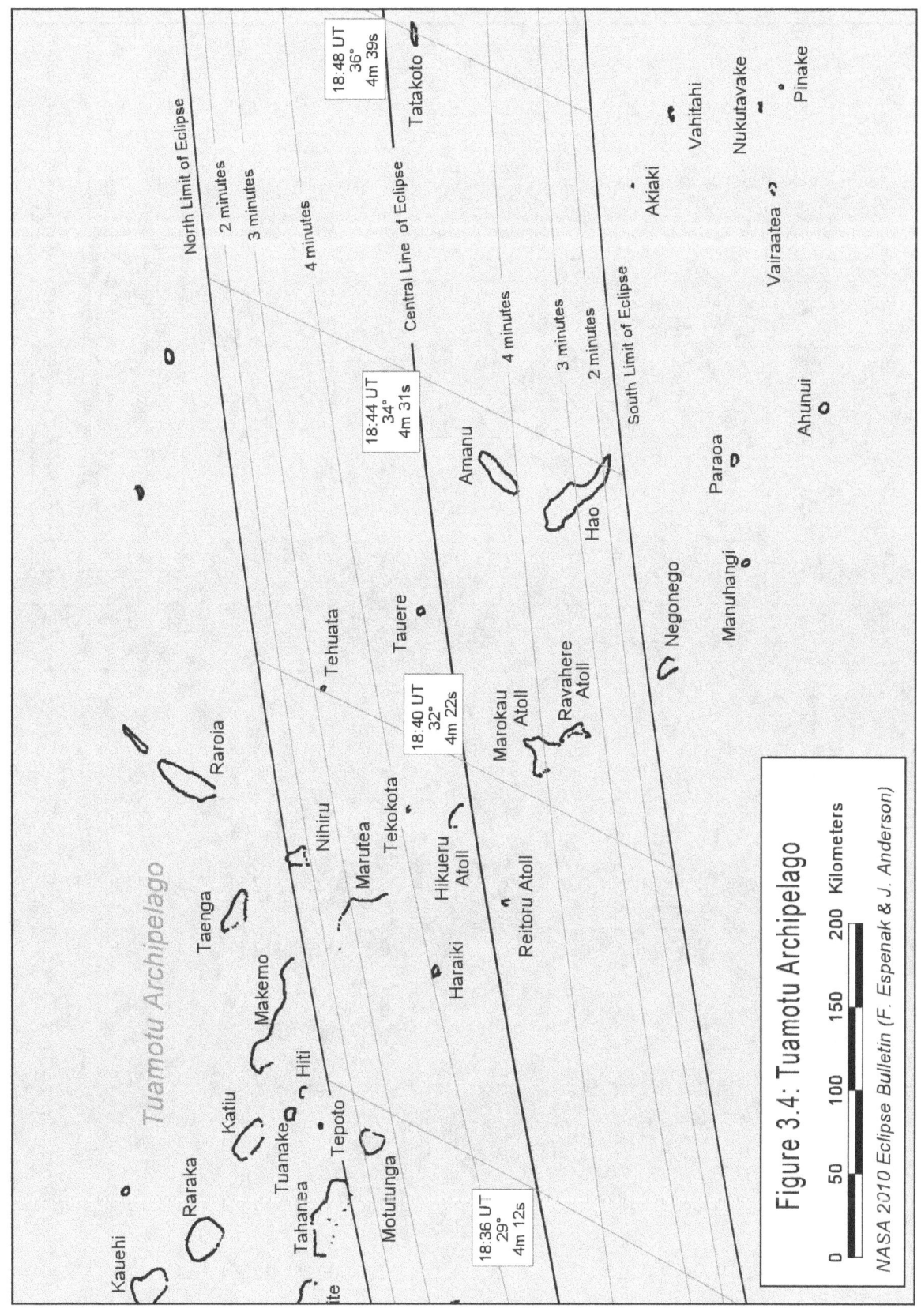

Figure 3.4: Tuamotu Archipelago

Total Solar Eclipse of 2010 July 11

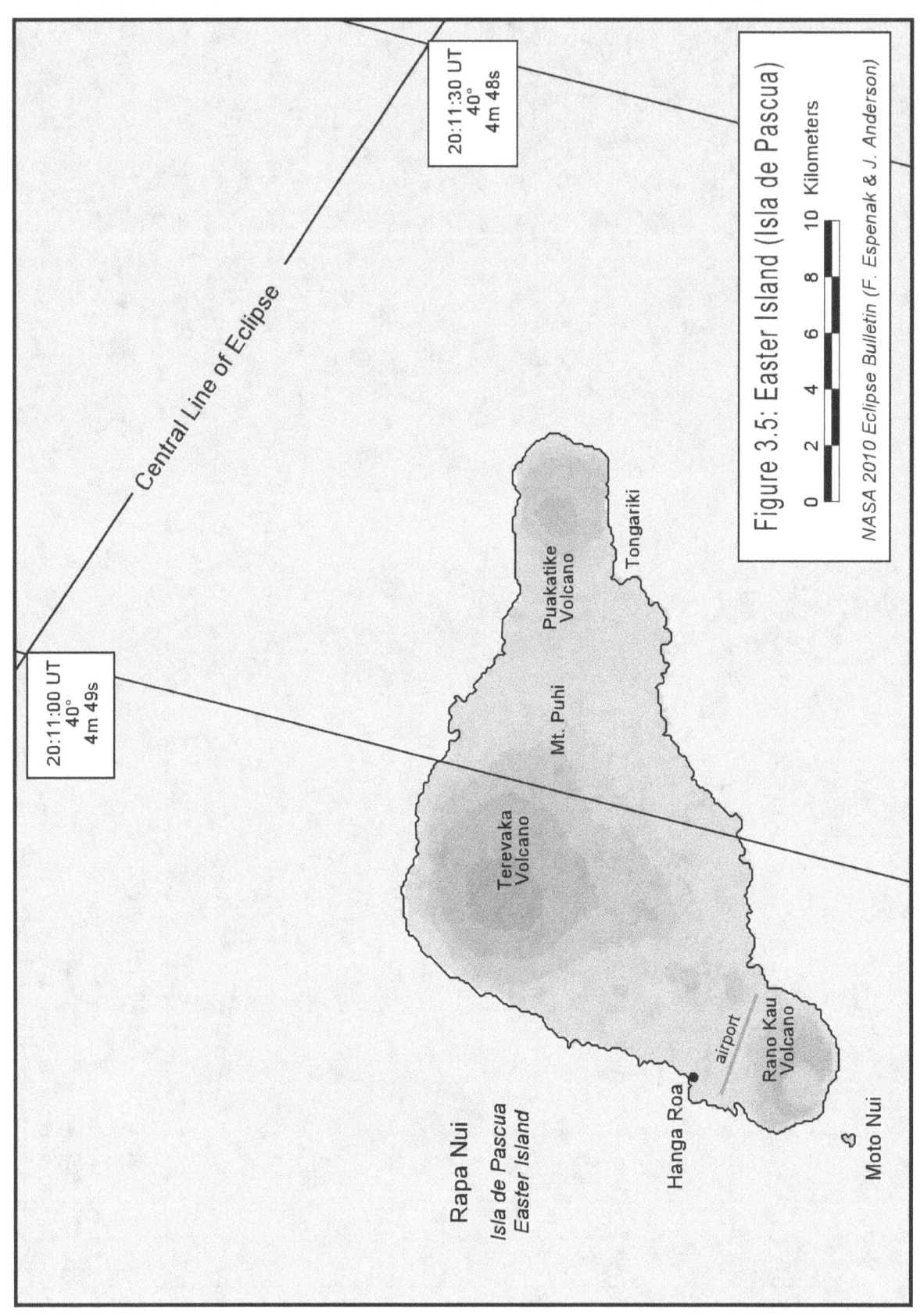

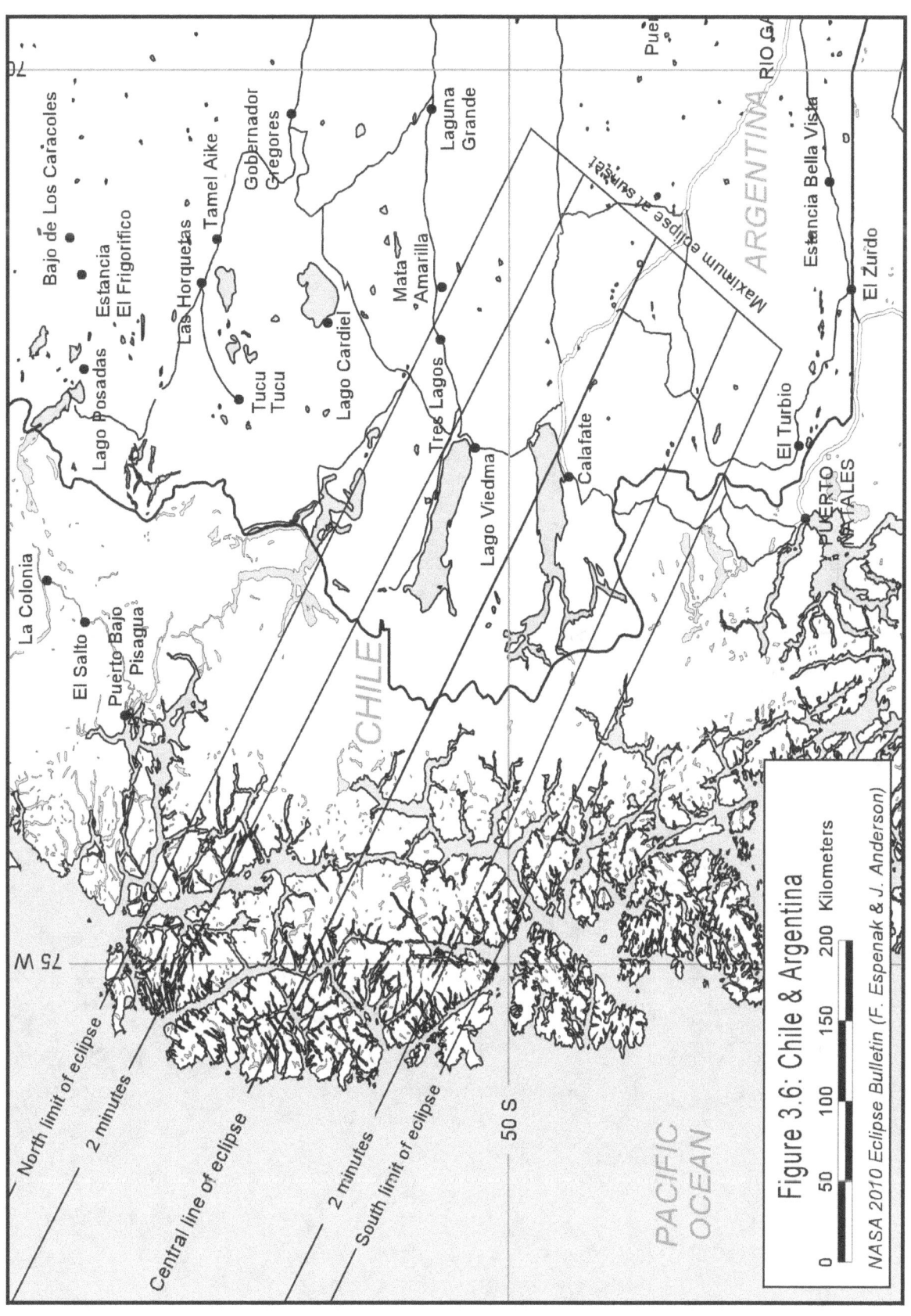

Figure 3.6: Chile & Argentina

Total Solar Eclipse of 2010 July 11

Figure 3.7: Lunar Limb Profile for July 11 at 19:30 UT

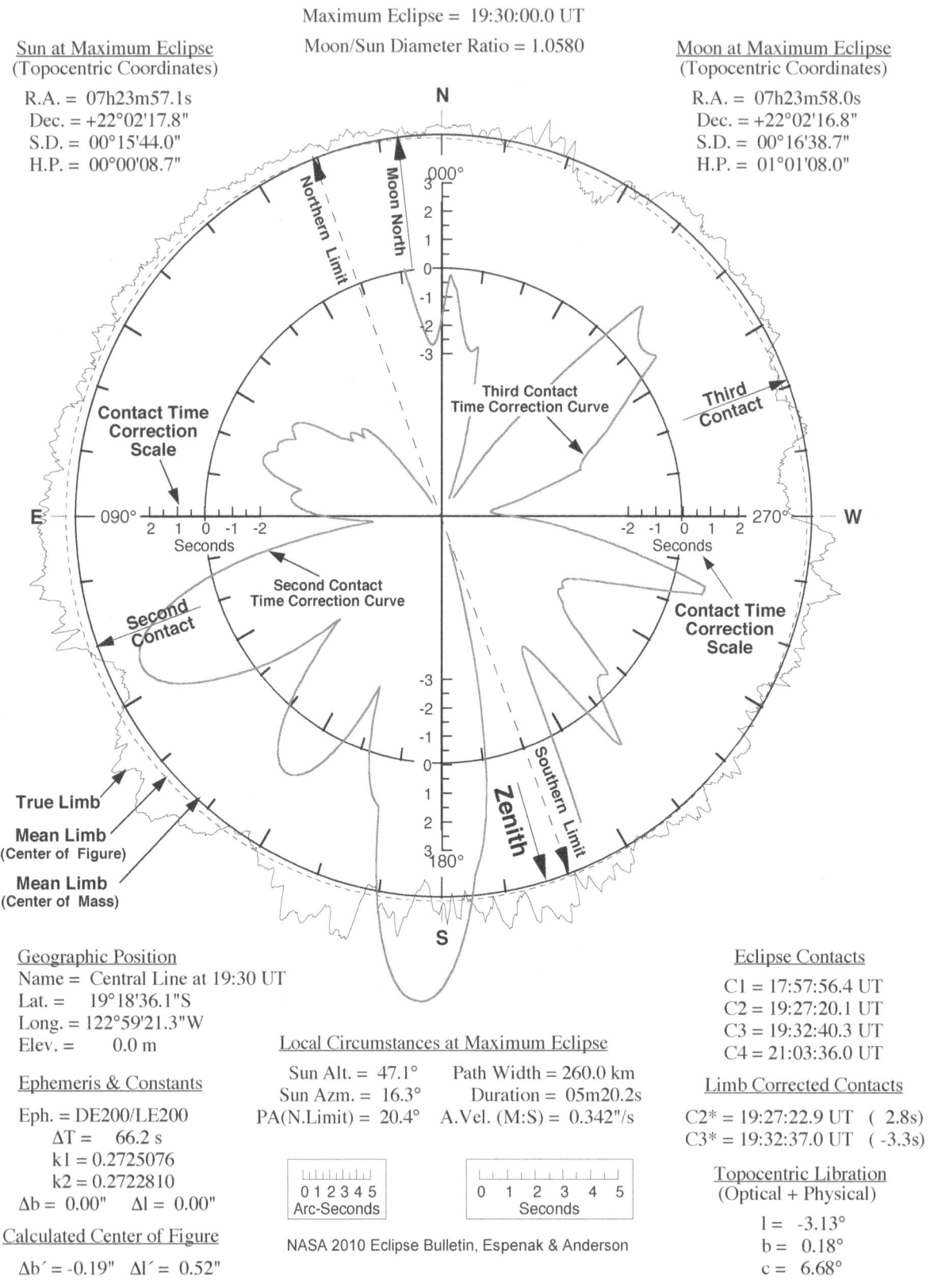

NASA 2010 Eclipse Bulletin, Espenak & Anderson

62 Annular and Total Solar Eclipses of 2010

Figure 3.8: Typical Weather Systems in July

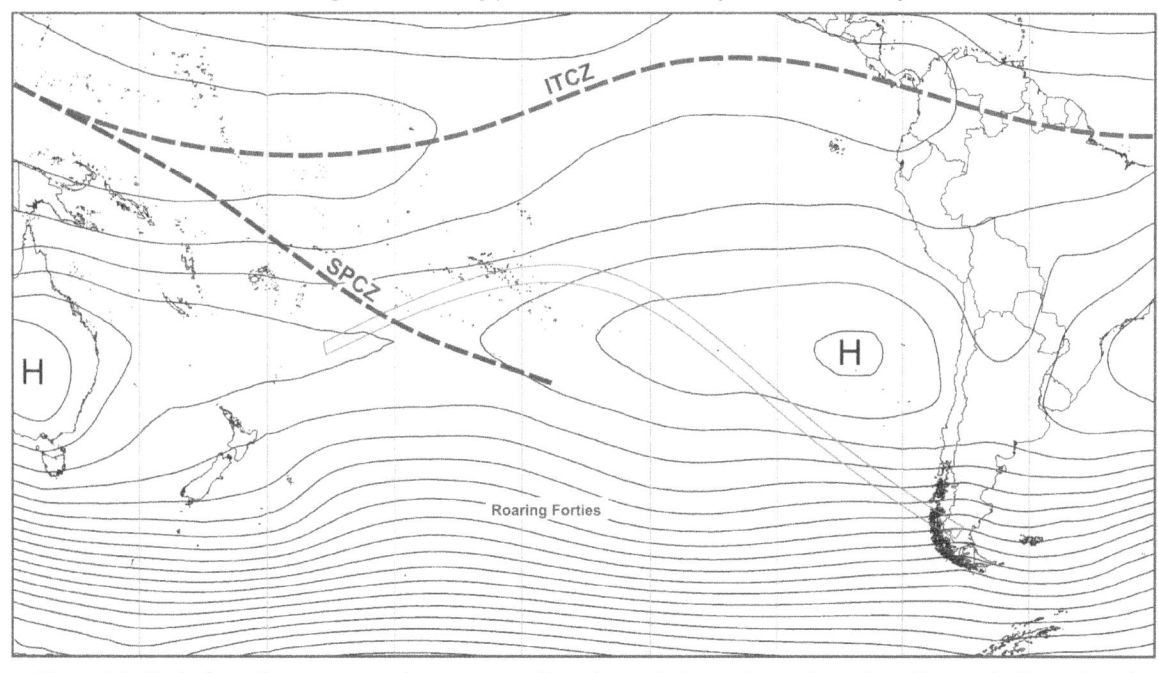

Figure 3.8: Typical weather systems and convergence lines during July are shown along the eclipse path. The string of high-pressure systems along the 30th parallel will have the most beneficial impact on the map of weather systems and convergence lines. Data courtesy Hadley Centre.

Figure 3.9: Average Cloudiness in July

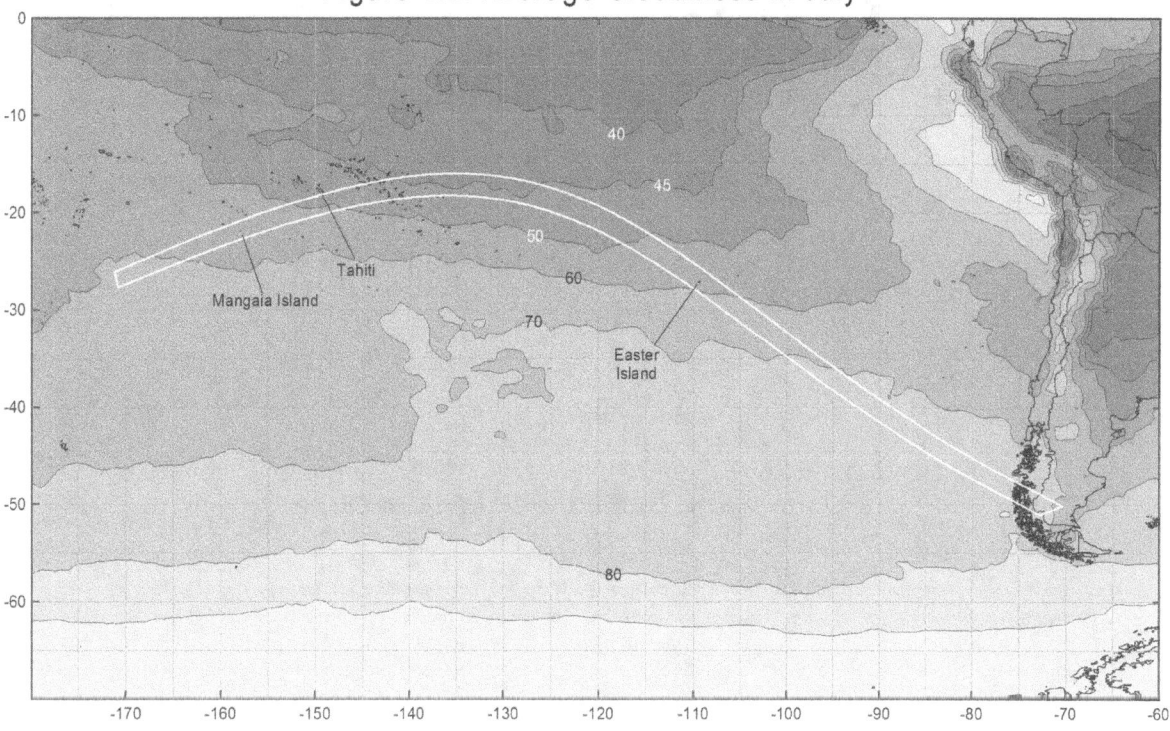

Figure 3.9: Average cloudiness in July in percent has been derived from 25 years of satellite imagery, up to 2004. This map is better suited to site-by-site comparison than the station data, as the biases inherent in the algorithms that collect the cloud statistics are better controlled than those associated with human observations at the individual stations.

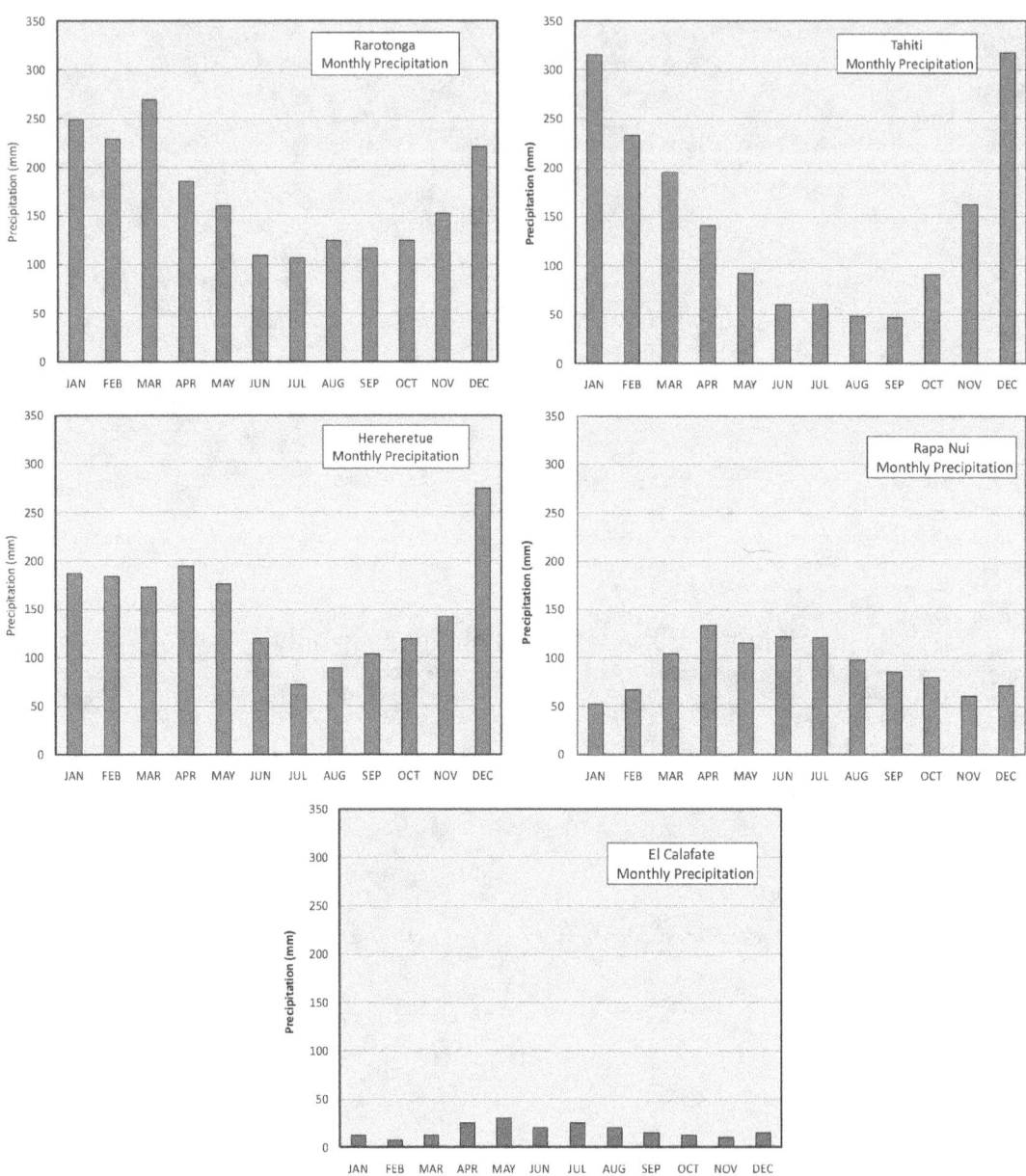

Figure 3.10: Annual precipitation graphs shown the monthly averages for five sites along the eclipse path. Note that western sites are in the midst of the seasonal dry spell, while Easter Island is in its wet season.

4. Observing Eclipses

4.1 Eye Safety and Solar Eclipses

A solar eclipse is probably the most spectacular astronomical event that many people will experience in their lives. There is a great deal of interest in watching eclipses, and thousands of astronomers (both amateur and professional) and other eclipse enthusiasts travel around the world to observe and photograph them.

A solar eclipse offers students a unique opportunity to see a natural phenomenon that illustrates the basic principles of mathematics and science taught through elementary and secondary school. Indeed, many scientists (including astronomers) have been inspired to study science as a result of seeing a total solar eclipse. Teachers can use eclipses to show how the laws of motion and the mathematics of orbits can predict the occurrence of eclipses. The use of pinhole cameras and telescopes or binoculars to observe an eclipse leads to an understanding of the optics of these devices. The rise and fall of environmental light levels during an eclipse illustrate the principles of radiometry and photometry, while biology classes can observe the associated behavior of plants and animals. It is also an opportunity for children of school age to contribute actively to scientific research—observations of contact timings at different locations along the eclipse path are useful in refining our knowledge of the orbital motions of the Moon and Earth, and sketches and photographs of the solar corona can be used to build a three-dimensional picture of the Sun's extended atmosphere during the eclipse.

Observing the Sun, however, can be dangerous if the proper precautions are not taken. The solar radiation that reaches the surface of Earth ranges from ultraviolet (UV) radiation at wavelengths longer than 290 nm, to radio waves in the meter range. The tissues in the eye transmit a substantial part of the radiation between 380–400 nm to the light-sensitive retina at the back of the eye. While environmental exposure to UV radiation is known to contribute to the accelerated aging of the outer layers of the eye and the development of cataracts, the primary concern over improper viewing of the Sun during an eclipse is the development of "eclipse blindness" or retinal burns.

Exposure of the retina to intense visible light causes damage to its light-sensitive rod and cone cells. The light triggers a series of complex chemical reactions within the cells which damages their ability to respond to a visual stimulus, and in extreme cases, can destroy them. The result is a loss of visual function, which may be either temporary or permanent depending on the severity of the damage. When a person looks repeatedly, or for a long time, at the Sun without proper eye protection, this photochemical retinal damage may be accompanied by a thermal injury—the high level of visible and near-infrared radiation causes heating that literally cooks the exposed tissue. This thermal injury or photocoagulation destroys the rods and cones, creating a small blind area. The danger to vision is significant because photic retinal injuries occur without any feeling of pain (the retina has no pain receptors), and the visual effects do not become apparent for at least several hours after the damage is done (Pitts 1993). Viewing the Sun through binoculars, a telescope, or other optical devices without proper protective filters can result in immediate thermal retinal injury because of the high irradiance level in the magnified image.

The only time that the Sun can be viewed safely with the naked eye is during a total eclipse, when the Moon completely covers the disk of the Sun. *It is never safe to look at a partial or annular eclipse, or the partial phases of a total solar eclipse, without the proper equipment and techniques.* Even when 99% of the Sun's surface (the photosphere) is obscured during the partial phases of a solar eclipse, the remaining crescent Sun is still intense enough to cause a retinal burn, even though illumination levels are comparable to twilight (Chou 1981 and 1996, and Marsh 1982). Failure to use proper observing methods may result in permanent eye damage and severe visual loss. This can have important adverse effects on career choices and earning potential, because it has been shown that most individuals who sustain eclipse-related eye injuries are children and young adults (Penner and McNair 1966, Chou and Krailo 1981, and Michaelides et al. 2001).

The same techniques for observing the Sun outside of eclipses are used to view and photograph annular solar eclipses and the partly eclipsed Sun (Sherrod 1981, Pasachoff 2000, Pasachoff and Covington 1993, and Reynolds and Sweetsir 1995). The safest and most inexpensive method is by projection. A pinhole or small opening is used to form an image of the Sun on a screen placed about a meter behind the opening. Multiple openings in perfboard, a loosely woven straw hat, or even interlaced fingers can be used to cast a pattern of solar images on a screen. A similar effect is seen on the ground below a broad-leafed tree: the many "pinholes" formed by overlapping leaves creates hundreds of crescent-shaped images. Binoculars or a small telescope mounted on a tripod can also be used to project a magnified image of the Sun onto a white card. All of these methods can be used to provide a safe view of the partial phases of an eclipse to a group of observers, but care must be taken to ensure that no one looks through the device. The main advantage of the projection methods is that nobody is looking directly at the Sun. The disadvantage of the pinhole method is that the screen must be placed at least a meter behind the opening to get a solar image that is large enough to be easily seen.

The Sun can only be viewed directly when filters specially designed to protect the eyes are used. Most of these filters have a thin layer of chromium alloy or aluminum deposited on their surfaces that attenuates both visible and near-infrared radiation. A safe solar filter should transmit less than 0.003% (density ~4.5) of visible light and no more than 0.5% (density ~2.3) of the near-infrared radiation from 780–1400 nm. (In addition to the term transmittance [in percent], the energy transmission of a filter can also be described by the term density [unit less] where density, d, is the common logarithm of the reciprocal of transmittance, t, or $d=\log 10[1/t]$. A density of '0' corresponds to a transmittance of 100%; a density of '1' corresponds to a transmittance of 10%; a density of '2' corresponds to a transmittance of 1%, etc.). Figure 4.1 shows transmittance curves for a selection of safe solar filters.

One of the most widely available filters for safe solar viewing is shade number 14 welder's glass, which can be obtained from welding supply outlets. A popular inexpensive alternative is aluminized polyester that has been specially made for solar observation. (This material is commonly known as "mylar," although the registered trademark "Mylar®" belongs to Dupont, which does not manufacture this material for use as a solar filter. Note that "space blankets" and aluminized polyester film used in gardening are NOT suitable for this purpose!) Unlike the welding glass, aluminized polyester can be cut to fit any viewing device, and does not break when dropped. It has been pointed out that some aluminized polyester filters may have large (up to approximately 1 mm in size) defects in their aluminum coatings that may be hazardous. A microscopic analysis of examples of such defects shows that despite their appearance, the defects arise from a hole in one of the two aluminized polyester films used in the filter. There is no large opening completely devoid of the protective aluminum coating. While this is a quality control problem, the presence of a defect in the aluminum coating does not necessarily imply that the filter is hazardous. When in doubt, an aluminized polyester solar filter that has coating defects larger than 0.2 mm in size, or more than a single defect in any 5 mm circular zone of the filter, should not be used.

An alternative to aluminized polyester that has become quite popular is "black polymer" in which carbon particles are suspended in a resin matrix. This material is somewhat stiffer than polyester film and requires a special holding cell if it is to be used at the front of binoculars, telephoto lenses, or telescopes. Intended mainly as a visual filter, the polymer gives a yellow-white image of the Sun (aluminized polyester produces a blue-white image). This type of filter may show significant variations in density of the tint across its extent; some areas may appear much lighter than others. Lighter areas of the filter transmit more infrared radiation than may be desirable. The advent of high resolution digital imaging in astronomy, especially for photographing the Sun, has increased the demand for solar filters of higher optical quality. Baader AstroSolar Safety Film, a metal-coated resin, can be used for both visual and photographic solar observations. A much thinner material, it has excellent optical quality and much less scattered light than polyester filters. The Baader material comes in two densities: one for visual use and a less dense version optimized for photography. Filters using optically flat glass substrates are available from several manufacturers, but are more expensive than polyester and polymer filters.

Many experienced solar observers use one or two layers of black-and-white film that has been fully exposed to light and developed to maximum density. Not all black-and-white films contain silver so care must be taken to use a silver-based emulsion. The metallic silver contained in the film acts as a protective filter; however, any black-and-white negative containing images is not suitable for this purpose. More recently, solar observers have used floppy disks and compact disks (CDs and CD-ROMs) as protective filters by covering the central openings and looking through the disk media. However, the optical quality of the solar image formed by a floppy disk or CD is relatively poor compared to aluminized polyester or welder's glass. Some CDs are made with very thin aluminum coatings that are not safe—if a lighted light bulb can be seen through the CD, it should not be used! No filter should be used with an optical device (e.g., binoculars, telescope, camera) unless it has been specifically designed for that purpose and is mounted at the front end. Some sources of solar filters are listed below.

Unsafe filters include color film, black-and-white film that contains no silver (i.e., chromogenic film), film negatives with images on them, smoked glass, sunglasses (single or multiple pairs), photographic neutral density filters and polarizing filters. Most of these transmit high levels of invisible infrared radiation, which can cause a thermal retinal burn (see Figure 4.1). The fact that the Sun appears dim, or that no discomfort is felt when looking at the Sun through the filter, is no guarantee that the eyes are safe.

Solar filters designed to thread into eyepieces that are often provided with inexpensive telescopes are also unsafe. These glass filters often crack unexpectedly from overheating when the telescope is pointed at the Sun, and retinal damage can occur faster than the observer can move the eye from the eyepiece. Avoid unnecessary risks. Local planetariums, science centers, or amateur astronomy clubs can provide additional information on how to observe the eclipse safely.

There are some concerns that ultraviolet-A (UVA) radiation (wavelengths from 315–380 nm) in sunlight may also adversely affect the retina (Del Priore 1999). While there is some experimental evidence for this, it only applies to the special case of aphakia, where the natural lens of the eye has been removed because of cataract or injury, and no UV-blocking spectacle, contact or intraocular lens has been fitted. In an intact normal human eye, UVA radiation does not reach the retina because it is absorbed by the crystalline lens. In aphakia, normal environmental exposure to solar UV radiation may indeed cause chronic retinal damage. The solar filter materials discussed in this article, however, attenuate solar UV radiation to a level well below the minimum permissible occupational exposure for UVA (ACGIH 2004), so an aphakic observer is at no additional risk of retinal damage when looking at the Sun through a proper solar filter.

In the days and weeks before a solar eclipse, there are often news stories and announcements in the media, warning about the dangers of looking at the eclipse. Unfortunately, despite the good intentions behind these messages, they frequently contain misinformation, and may be designed to scare people from viewing the eclipse at all. This tactic may backfire, however, particularly when the messages are intended for students. A student who heeds warnings from teachers and other authorities not to view the eclipse because of the danger to vision, and later learns that other students did see it safely, may feel cheated out of the experience. Having now learned that the authority figure was wrong on one occasion, how is this student going to react when other health-related advice about drugs, AIDS[2], or smoking is given (Pasachoff 2001). Misinformation may be just as bad, if not worse, than no information.

Remember that the total phase of an eclipse can, and should, be seen without any filters, and certainly never by projection! It is completely safe to do so. Even after observing 14 solar eclipses, the author finds the naked-eye view of the *totally eclipsed* Sun awe-inspiring. The experience should be enjoyed by all.

> Sect. 4.1 was contributed by:
> B. Ralph Chou, MSc, OD
> Associate Professor, School of Optometry
> University of Waterloo
> Waterloo, Ontario, Canada N2L 3G1

4.2 Sources for Solar Filters

The following is a brief list of sources for filters that are specifically designed for safe solar viewing with or without a telescope. The list is not meant to be exhaustive, but is a representative sample of sources for solar filters currently available in North America and Europe. For additional sources, see advertisements in *Astronomy* and or *Sky & Telescope* magazines. (The inclusion of any source on the following list does not imply an endorsement of that source by either the authors or NASA.)

Sources in the USA:

American Paper Optics, 3080 Bartlett Corporate Drive, Bartlett, TN 38133, (800) 767-8427 or (901) 381-1515

Astro-Physics, Inc., 11250 Forest Hills Rd., Rockford, IL 61115, (815) 282-1513.

Celestron International, 2835 Columbia Street, Torrance, CA 90503, (310) 328-9560.

Coronado Technology Group, 1674 S. Research Loop, Suite 436, Tucson, AZ 85710-6739, (520) 760-1561, (866) SUNWATCH.

DayStar Filters LLC, 149 Northwest OO Highway, Warrensburg, MO 64093, (660) 747-2100.

Meade Instruments Corporation, 16542 Millikan Ave., Irvine, CA 92606, (714) 756-2291.

Rainbow Symphony, Inc., 6860 Canby Ave., #120, Reseda, CA 91335, (818) 708-8400.

Telescope and Binocular Center, P.O. Box 1815, Santa Cruz, CA 95061-1815, (408) 763-7030.

Thousand Oaks Optical, Box 4813, Thousand Oaks, CA 91359, (805) 491-3642.

Sources in Canada:

Kendrick Astro Instruments, 2920 Dundas St. W., Toronto, Ontario, Canada M6P 1Y8, (416) 762-7946.

Khan Scope Centre, 3243 Dufferin Street, Toronto, Ontario, Canada M6A 2T2, (416) 783-4140.

Perceptor Telescopes TransCanada, Brownsville Junction Plaza, Box 38, Schomberg, Ontario, Canada L0G 1T0, (905) 939-2313.

Sources in Europe:

Baader Planetarium GmbH, Zur Sternwarte, 82291 Mammendorf, Germany, 0049 (8145) 8802.

4.3 Eclipse Photography

A solar eclipse can be safely photographed provided the above precautions are followed. Almost any kind of camera can be used to capture this rare event, but Single Lens Reflex (SLR) cameras offer interchangeable lenses and zooms. A lens with a fairly long focal length is recommended in order to produce as large an image of the Sun as possible. A standard 50 mm lens on a 35 mm film camera yields a minuscule 0.5 mm solar image, while a 200 mm telephoto or zoom lens produces a 1.9 mm image (Figure 4.2). A better choice would be one of the small, compact, catadioptic or mirror lenses that have become widely available in the past 20 years. The focal length of 500 mm is most common among such mirror lenses and yields a solar image of 4.6 mm.

With one solar radius of corona on either side, an eclipse view during totality will cover 9.2 mm. Adding a 2x teleconverter will produce a 1000 mm focal length, which doubles the Sun's diameter to 9.2 mm. Focal lengths in excess of 1000 mm usually fall within the realm of amateur telescopes.

Consumer digital cameras have become affordable in recent years and many of these may be used to photograph the eclipse. Most recommendations for 35 mm SLR cameras apply to digital SLR (DSLR) cameras as well. The primary difference is that the imaging chip in most DSLR cameras is only about 2/3 the area of a 35 mm film frame (check the camera's technical specifications). This means that the Sun's relative size will be 1.5 times larger in a DSLR camera so a shorter focal length lens can be used to achieve the same angular coverage compared to a 35 mm SLR camera. For example, a 500 mm lens on a digital camera produces the same relative image size as a 750 mm lens on a 35 mm camera (Figure 4.2). Another issue to consider is the lag time between digital frames required to write images to the DSLR's memory card. Better DSLRs have a buffer to temporarily store a burst of images before they are written to the card. It is also advisable to turn off the autofocus because it is not reliable under these conditions; focus the camera manually instead. Preparations must also be made for adequate battery power and space on the memory card.

If full disk photography of an annular or partial eclipse is planned, the focal length of the optics must not exceed 2500 mm on 35 mm format (1700 mm on digital). Longer focal lengths permit photography of only a magnified portion of the Sun's disk. In order to photograph the Sun's corona during a total eclipse, the focal length should be no longer than about 1500 mm (1000 mm on digital); however, a shorter focal length of 1000 mm (700 mm digital) requires less critical framing and can capture some of the longer coronal streamers. Figure 4.2

---

2. Acquired Immunodeficiency Syndrome

shows the apparent size of the Sun (or Moon) and the outer corona in both film and digital formats for a range of lens focal lengths. For any particular focal length, the diameter of the Sun's image (on 35 mm film) is approximately equal to the focal length divided by 109 (Table 4.1).

A solar filter must be used on the lens throughout the partial phases for both photography and safe viewing. Such filters are most easily obtained through manufacturers and dealers listed in *Sky & Telescope* and *Astronomy* magazines (see Sect. 4.2, "Sources for Solar Filters"). These filters typically attenuate the Sun's visible and infrared energy by a factor of 100,000. The actual filter factor and choice of International Organization for Standardization (ISO) speed, however, will play critical roles in determining the correct photographic exposure. Almost any ISO can be used because the Sun gives off abundant light. The easiest method for determining the correct exposure is accomplished by running a calibration test on the uneclipsed Sun. Shoot a roll of film of the mid-day Sun at a fixed aperture (f/8 to f/16) using every shutter speed from 1/1000 s to 1/4 s. After the film is developed, note the best exposures and use them to photograph all the partial phases. With a digital camera, the process is even easier: shoot a range of different exposures and use the camera's histogram display to evaluate the best exposure. The Sun's surface brightness remains constant throughout the eclipse, so no exposure compensation is needed except for the narrow crescent phases, which require two more stops due to solar limb darkening. Bracketing by several stops is also necessary if haze or clouds interfere on eclipse day.

Certainly the most spectacular and awe-inspiring phase of an eclipse is totality. For a few brief minutes or seconds, the Sun's pearly white corona, red prominences, and chromosphere are visible. The great challenge is to obtain a set of photographs that captures these fleeting phenomena. The most important point to remember is that during the total phase, all solar filters must be removed. The corona has a surface brightness a million times fainter than the photosphere, so photographs of the corona must be made *without* a filter. Furthermore, it is completely safe to view the totally eclipsed Sun directly with the naked eye. No filters are needed, and in fact, they would only hinder the view. The average brightness of the corona varies inversely with the distance from the Sun's limb. The inner corona is far brighter than the outer corona so no single exposure can capture its full dynamic range. The best strategy is to choose one aperture or f/number and bracket the exposures over a range of shutter speeds (e.g., 1/1000 s to 1 s). Rehearsing this sequence is highly recommended because great excitement accompanies totality and there is little time to think.

Exposure times for various combinations of ISO speeds, apertures (f/number) and solar features (chromosphere, prominences, inner, middle, and outer corona) are summarized in Table 4.2. The table was developed from eclipse photographs made by F. Espenak, as well as from photographs published in *Sky and Telescope*. To use the table, first select the ISO speed in the upper left column. Next, move to the right to the desired aperture or f/number for the chosen ISO speed. The shutter speeds in that column may be used as starting points for photographing various features and phenomena tabulated in the 'Subject' column at the far left. For example, to photograph prominences using ISO 400 at f/16, the table recommends an exposure of 1/1000. Alternatively, the recommended shutter speed can be calculated using the 'Q' factors tabulated along with the exposure formula at the bottom of Table 4.2. Keep in mind that these exposures are based on a clear sky and a corona of average brightness. The exposures should be bracketed one or more stops to take into account the actual sky conditions and the variable nature of these phenomena.

Point-and-shoot cameras with wide angle lenses are excellent for capturing the quickly changing light in the seconds before and during totality. Use a tripod or brace with the camera on a wall or fence because slow shutter speeds will be needed. In addition, disable or turn off the camera's electronic flash so that it does not interfere with anyone else's view of the eclipse. If the flash cannot be turned off, cover it with black tape.

Another eclipse effect that is easily captured with point-and-shoot cameras should not be overlooked. Use a straw hat or a kitchen sieve and allow its shadow to fall on a piece of white cardboard placed several feet away. The small holes act like pinhole cameras and each one projects its own image of the eclipsed Sun. The effect can also be duplicated by forming a small aperture with the fingers of one's hands and watching the ground below. The pinhole camera effect becomes more prominent with increasing eclipse magnitude. Virtually any camera can be used to photograph the phenomenon, but automatic cameras must have their flashes turned off because this would otherwise obliterate the pinhole images.

For more information on eclipse photography, observations, and eye safety, see the "Further Reading" sections in the References.

4.4 Contact Timings from the Path Limits

Precise timings of beading phenomena made near the northern and southern limits of the umbral path (i.e., the graze zones), may be useful in determining the diameter of the Sun relative to the Moon at the time of the eclipse. Such measurements are essential to an ongoing project to try to detect changes in the solar diameter.

Because of the conspicuous nature of the eclipse phenomena and their strong dependence on geographical location, scientifically useful observations can be made with relatively modest equipment. A small telescope of 3- to 5-in (75–125 mm) aperture, portable shortwave radio, and portable camcorder comprise standard equipment used to make such measurements. Time signals are broadcast via shortwave stations such as WWV and CHU in North America (5.0, 10.0, 15.0, and 20.0 MHz are example frequencies to try for these signals around the world), and are recorded simultaneously as the eclipse is videotaped. Those using video are encouraged to use one of the Global Positioning System (GPS) video time inserters, such as the Kiwi OSD by PFD systems (http://www.pfdsystems.com) in order to link specific Baily's bead events with lunar features.

The safest timing technique consists of observing a projection of the Sun rather than directly imaging the solar disk itself. If a video camera is not available, a tape recorder can be used to record time signals with verbal timings of each event. Inexperienced observers are cautioned to use great care in making such observations.

The method of contact timing should be described in detail, along with an estimate of the error. The precision requirements of these observations are ±0.5 s in time, 1 arcsec (~30 m) in latitude and longitude, and ±20 m (~60 ft) in elevation. Commercially available GPS receivers are now the easiest and best way to determine one's position to the necessary accuracy. GPS receivers are also a useful source for accurate Universal Time as long as they use the one-pulse-per-second signal for timing; many receivers do not use that, so the receiver's specifications must be checked. The National Marine Electronics Association (NMEA) sequence normally used can have errors in the time display of several tenths of a second.

The observer's geodetic coordinates are best determined with a GPS receiver. Even simple handheld models are fine if data are obtained and averaged until the latitude, longitude, and altitude output become stable. Positions can also be measured from United States Geological Survey (USGS) maps or other large scale maps as long as they conform to the accuracy requirement above. Some of these maps are available on Web sites such as <http://www.topozone.co>. Coordinates determined directly from Web sites are useful for checking, but are usually not accurate enough for eclipse timings. If a map or GPS is unavailable, then a detailed description of the observing site should be included, providing information such as distance and directions of the nearest towns or settlements, nearby landmarks, identifiable buildings, and road intersections; digital photos of key annotated landmarks are also important.

Expeditions are coordinated by the International Occultation Timing Association (IOTA). For information on possible solar eclipse expeditions that focus on observing at the eclipse path limits, refer to <http://www.eclipsetours.com>. For specific details on equipment and observing methods for observing at the eclipse path limits, refer to <http://www.eclipsetours.com/edge>. For more information on IOTA and eclipse timings, contact:

Dr. David W. Dunham, IOTA
Johns Hopkins University/Applied Physics Lab.
MS MP3-135
11100 Johns Hopkins Rd.
Laurel, MD 20723–6099, USA
Phone: (240) 228-5609
E-mail: dunham@starpower.net
Web Site: http://www.lunar-occultations.com/iota

Reports containing graze observations, eclipse contact, and Baily's bead timings, including those made anywhere near, or in, the path of totality or annularity can be sent to Dr. Dunham at the address listed above.

4.5 Plotting Eclipse Paths on Maps

To assist hand-plotting of high-resolution maps of the umbral path, high resolution tables of graze zone coordinates at longitude increments of 7.5′ are available via the NASA Web sites for the 2010 annular eclipse <http://eclipse.gsfc.nasa.gov/SEmono/ASE2010/ASE2010.html>, and the 2010 total eclipse <http://eclipse.gsfc.nasa.gov/SEmono/TSE2010/TSE2010.html>.

Global Navigation Charts (1:5,000,000), Operational Navigation Charts (scale 1:1,000,000), and Tactical Pilotage Charts (1:500,000) of the world are published by the National Imagery and Mapping Agency. Sales and distribution of these maps are through the National Ocean Service. For specific information about map availability, purchase prices, and ordering instructions, the National Ocean Service can be contacted by mail, telephone, or fax at the following:

NOAA Distribution Division, N/ACC3
National Ocean Service
Riverdale, MD 20737–1199, USA
Phone: (301) 436-8301 or (800) 638-8972
Fax: (301) 436-6829

It is also advisable to check the telephone directory for any map specialty stores in a given city or area. They often have large inventories of maps available for immediate delivery.

4.6 Eclipse Paths on Google Maps

The 2010 eclipse paths are also plotted on interactive Google Maps on the NASA Eclipse Web site. The northern and southern path limits of an eclipse path are plotted in blue and the central lines are red. The yellow lines crossing the path indicate the position of maximum eclipse at 10 min intervals. The four-way toggle arrows (upper left corner) are for navigating around the map. The zoom bar (left edge) is used to change the magnification. The three buttons (top right) turn on the map view, the satellite view, or the hybrid map/satellite view.

The green marker labeled GE is the point of Greatest Eclipse. Clicking anywhere on a map marks a position and calculates the eclipse times at that location. Moving the cursor over a marker reveals the eclipse circumstances for that position. The marker predictions can also be viewed in a new window via the *Eclipse Times Popup* button. The information in the popup window can be selected, copied, and pasted into a word processor. All markers can be removed using the *Clear Markers* button above. Choosing the *Large Map* check box produces a bigger map (for users with large monitors and fast Internet connections).

The URL for the Google Map of the 2010 annular eclipse is <http://eclipse.gsfc.nasa.gov/SEgoogle/SEgoogle2001/SE-2010Jan15Agoogle2.html>. The URL for Google Map of the 2010 total eclipse is <http://eclipse.gsfc.nasa.gov/SEgoogle/SEgoogle2001/SE2010Jul11Tgoogle2.html>.

Table 4.1:
Field of View and Size of Sun's Image for Various Photographic Focal Lengths

Focal Length	Field of View (35mm)	Field of View (digital)	Size of Sun
14 mm	98° x 147°	65° x 98°	0.2 mm
20 mm	69° x 103°	46° x 69°	0.2 mm
28 mm	49° x 74°	33° x 49°	0.2 mm
35 mm	39° x 59°	26° x 39°	0.3 mm
50 mm	27° x 40°	18° x 28°	0.5 mm
105 mm	13° x 19°	9° x 13°	1.0 mm
200 mm	7° x 10°	5° x 7°	1.8 mm
400 mm	3.4° x 5.1°	2.3° x 3.4°	3.7 mm
500 mm	2.7° x 4.1°	1.8° x 2.8°	4.6 mm
1000 mm	1.4° x 2.1°	0.9° x 1.4°	9.2 mm
1500 mm	0.9° x 1.4°	0.6° x 0.9°	13.8 mm
2000 mm	0.7° x 1.0°	0.5° x 0.7°	18.4 mm

Image Size of Sun (mm) = Focal Length (mm) / 109

Table 4.2: Solar Eclipse Exposure Guide

ISO				f/Number				
25	1.4	2	2.8	4	5.6	8	11	16	22
50	2	2.8	4	5.6	8	11	16	22	32
100	2.8	4	5.6	8	11	16	22	32	44
200	4	5.6	8	11	16	22	32	44	64
400	5.6	8	11	16	22	32	44	64	88
800	8	11	16	22	32	44	64	88	128
1600	11	16	22	32	44	64	88	128	176

Subject	Q				Shutter Speed				
Solar Eclipse									
Partial[1] - 4.0 ND	11	—	—	—	1/4000	1/2000	1/1000	1/500	1/250	1/125
Partial[1] - 5.0 ND	8	1/4000	1/2000	1/1000	1/500	1/250	1/125	1/60	1/30	1/15
Baily's Beads[2]	11	—	—	—	1/4000	1/2000	1/1000	1/500	1/250	1/125
Chromosphere	10	—	—	1/4000	1/2000	1/1000	1/500	1/250	1/125	1/60
Prominences	9	—	1/4000	1/2000	1/1000	1/500	1/250	1/125	1/60	1/30
Corona - 0.1 Rs	7	1/2000	1/1000	1/500	1/250	1/125	1/60	1/30	1/15	1/8
Corona - 0.2 Rs[3]	5	1/500	1/250	1/125	1/60	1/30	1/15	1/8	1/4	1/2
Corona - 0.5 Rs	3	1/125	1/60	1/30	1/15	1/8	1/4	1/2	1 sec	2 sec
Corona - 1.0 Rs	1	1/30	1/15	1/8	1/4	1/2	1 sec	2 sec	4 sec	8 sec
Corona - 2.0 Rs	0	1/15	1/8	1/4	1/2	1 sec	2 sec	4 sec	8 sec	15 sec
Corona - 4.0 Rs	-1	1/8	1/4	1/2	1 sec	2 sec	4 sec	8 sec	15 sec	30 sec
Corona - 8.0 Rs	-3	1/2	1 sec	2 sec	4 sec	8 sec	15 sec	30 sec	1 min	2 min

Exposure Formula: $t = f^2 / (I \times 2^Q)$ where: t = exposure time (sec)
f = f/number or focal ratio
I = ISO film speed
Q = brightness exponent

Abbreviations: ND = Neutral Density Filter.
Rs = Solar Radii.

Notes: [1] Exposures for partial phases are also good for annular eclipses.
[2] Baily's Beads are extremely bright and change rapidly.
[3] This exposure also recommended for the 'Diamond Ring' effect.

F. Espenak - 2006 Oct

FIGURE 4.1: SPECTRAL RESPONSE OF SOME COMMONLY AVAILABLE SOLAR FILTERS

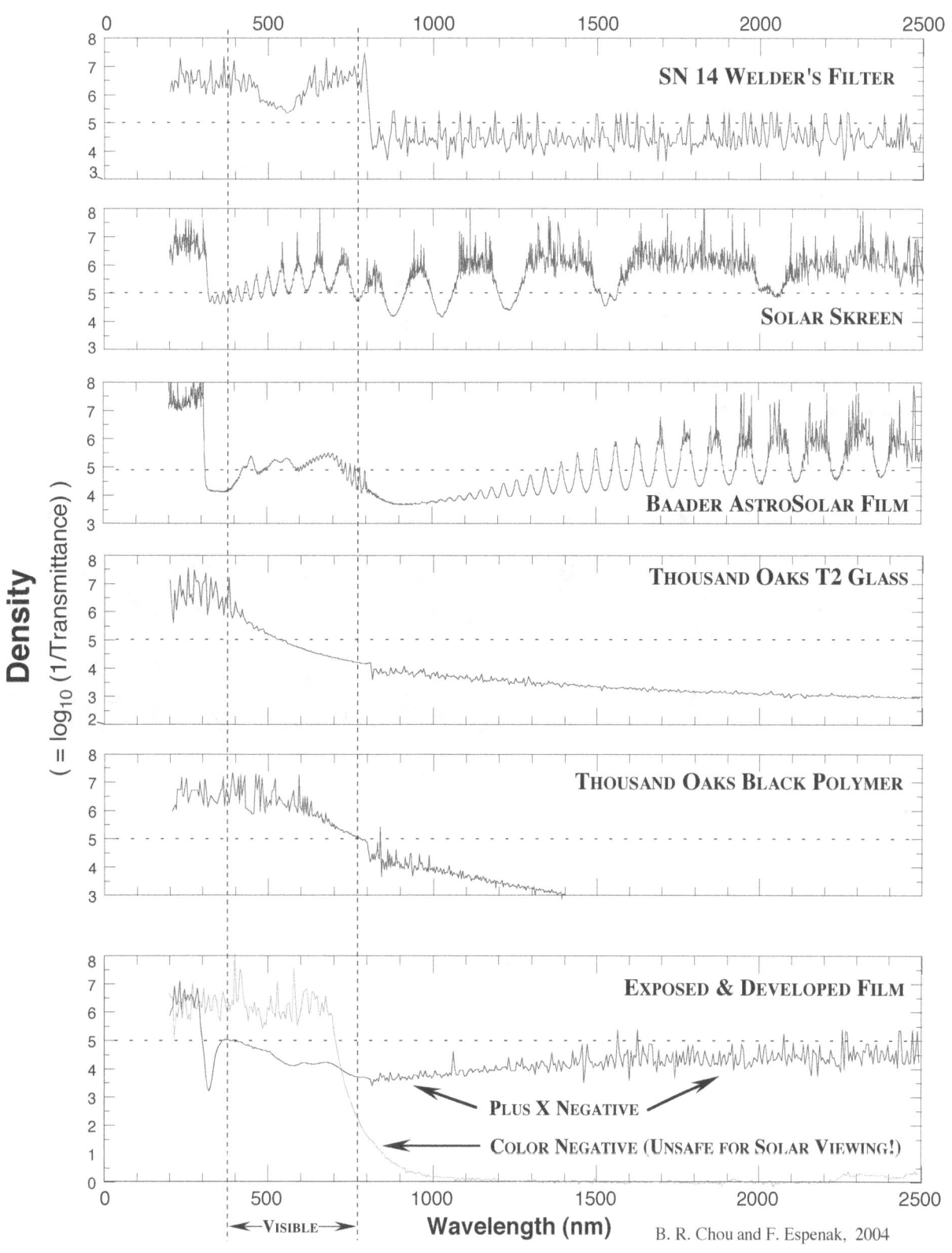

Figure 4.2 - Lens Focal Length vs. Image Size for Eclipse Photography

The image size of the eclipsed Sun and corona is shown for a range of focal lengths on both 35 mm film cameras and digital SLR's which use a CCD 2/3 the size of 35 mm film. Thus, the same lens produces an image 1.5 x larger on a digital SLR as compared to film.

5. Eclipse Resources

5.1 IAU Working Group on Eclipses

Professional scientists are asked to send descriptions of their eclipse plans to the Working Group on Eclipses of the Solar Division of the International Astronomical Union (IAU), so they can keep a list of observations planned. Send such descriptions, even in preliminary form, to:

International Astronomical Union/
    Working Group on Eclipses
Prof. Jay M. Pasachoff, Chair
Williams College–Hopkins Observatory
Williamstown, MA 01267, USA
Fax: (413) 597-3200
E-mail: eclipse@williams.edu
Web Site: http://www.eclipses.info

The members of the Working Group on Eclipses of the Solar Division of the IAU are: Jay M. Pasachoff (USA), Chair, Iraida S. Kim (Russia), Hiroki Kurokawa (Japan), Jagdev Singh (India), Vojtech Rusin (Slovakia), Fred Espenak (USA), Jay Anderson (Canada), Glenn Schneider (USA), Michael Gill (UK), and Yihua Yan (China).

5.2 IAU Solar Eclipse Education Committee

In order to ensure that astronomers and public health authorities have access to information on safe viewing practices, the Commission on Education and Development of the IAU, set up a Program Group on Public Education at the Times of Eclipses. Under Prof. Jay M. Pasachoff, the Committee has assembled information on safe methods of observing solar eclipses, eclipse-related eye injuries, and samples of educational materials on solar eclipses (see <http://www.eclipses.info>).

For more information, contact Prof. Jay M. Pasachoff (contact information is found in Sect. 5.1). Information on safe solar filters can be obtained by contacting Program Group member Dr. B. Ralph Chou (bchou@sciborg.uwaterloo.ca).

5.3 Solar Eclipse Mailing List

The Solar Eclipse Mailing List (SEML) is an electronic news group dedicated to solar eclipses. Published by British eclipse chaser Michael Gill (eclipsechaser@yahoo.com), it serves as a forum for discussing anything and everything about eclipses and facilitates interaction between both the professional and amateur communities.

The SEML is hosted at URL <http://groups.yahoo.com/group/SEML/>. Complete instructions are available online for subscribing and unsubscribing. Up until mid-2004, the list manager of the SEML was Patrick Poitevin (solareclipsewebpages@btopenworld.com). Archives of past SEML messages through July 2004 are available at <http://www.mreclipse.com/SENL/SENLinde.htm>.

5.4 NASA Eclipse Bulletins on the Internet

To make the NASA solar eclipse bulletins accessible to as large an audience as possible, these publications are also available via the Internet. The bulletins can be read, or downloaded using a Web browser (Firefox, Safari, Internet Explorer, etc.) from the NASA Eclipse Web Site. The top-level Web addresses (URLs) for the currently available eclipse bulletins are as follows:

Annular Solar Eclipse of 1994 May 10
— http://eclipse.gsfc.nasa.gov/SEpubs/19940510/rp.html
Total Solar Eclipse of 1994 Nov 03
— http://eclipse.gsfc.nasa.gov/SEpubs/19941103/rp.html
Total Solar Eclipse of 1995 Oct 24
— http://eclipse.gsfc.nasa.gov/SEpubs/19951024/rp.html
Total Solar Eclipse of 1997 Mar 09
— http://eclipse.gsfc.nasa.gov/SEpubs/19970309/rp.html
Total Solar Eclipse of 1998 Feb 26
— http://eclipse.gsfc.nasa.gov/SEpubs/19980226/rp.html
Total Solar Eclipse of 1999 Aug 11
— http://eclipse.gsfc.nasa.gov/SEpubs/19990811/rp.html
Total Solar Eclipse of 2001 Jun 21
— http://eclipse.gsfc.nasa.gov/SEpubs/20010621/rp.html
Total Solar Eclipse of 2002 Dec 04
— http://eclipse.gsfc.nasa.gov/SEpubs/20021204/rp.html
Solar Eclipses of 2003: May 31 & Nov 23
— http://eclipse.gsfc.nasa.gov/SEpubs/20030000/rp.html
Total Solar Eclipse of 2006 Mar 29
— http://eclipse.gsfc.nasa.gov/SEpubs/20060329/rp.html
Total Solar Eclipse of 2008 Aug 01
— http://eclipse.gsfc.nasa.gov/SEpubs/20080801/rp.html
Total Solar Eclipse of 2009 Jul 22
— http://eclipse.gsfc.nasa.gov/SEpubs/20090722/rp.html
Annular and Total Solar Eclipses of 2010
— http://eclipse.gsfc.nasa.gov/SEpubs/2010/rp.html

The most recent bulletins are available in "PDF" format. All future NASA eclipse bulletins will be available over the Internet, at or before publication of each. Comments and suggestions are actively solicited to fix problems and improve on compatibility and formats..

5.5 Future Eclipse Paths on the Internet

Presently, the NASA eclipse bulletins are published 12–24 months before each eclipse, however, there have been a growing number of requests for eclipse path data with an even greater lead time. To accommodate this need, predictions have been generated for all central solar eclipses from 1901 through 2100. The umbral path characteristics have been calculated with a 1 min time interval compared to the 6 min interval used in *Fifty Year Canon of Solar Eclipses: 1986–2035* (Espenak 1987). This provides enough detail for making preliminary plots of the path on larger scale maps. Links to global maps using an orthographic projection present the regions of partial and total (or annular) eclipse. There are also small animations that show the motion of the umbral and penumbral shadows across Earth for each eclipse. To present all this information, a series of Web pages break the 200 year period into decade-long intervals. The Web page for the decade 2001–2010 is: <http://eclipse.gsfc.nasa.gov/SEcat/SEdecade2001.html>. Links to the other decades can be found on this page as well.

Google Maps is an excellent tool for a detailed look at past and future eclipse paths. A series of Google Maps Web pages has been created for all central eclipses from 1901–2100. The indices and links for these maps are arranged in 20-year periods. For example, the Web page for the period 2001–2020 is: <http://eclipse.gsfc.nasa.gov/SEgoogle/SEgoogle2001.html>. Links to the other 20-year index pages also can be found on this page.

A Web-based search engine has been developed with the assistance of Xavier Jubier and Sumit Dutta. It accesses the entire catalog of Besselian elements used in *Five Millennium Canon of Solar Eclipses: –1999 to +3000* (Espenak and Meeus 2006). The user can search this data by eclipse type, duration, and date range. The resulting table has links to coordinate tables and eclipse paths plotted on Google Maps. The link for *Five Millennium Solar Eclipse Search Engine* is: <http://eclipse.gsfc.nasa.gov/SEsearch/SEsearch.php>.

The coordinates of the Sun used in these tables and maps were calculated on the basis of the VSOP87 theory constructed by Bretagnon and Francou (1988). The Moon ephemeris is based on the theory ELP-2000/82 of Chapront-Touze and Chapront (1983). Neglecting the smallest periodic terms, the Moon's position calculated in our program has a mean error (as compared to the full ELP theory) of about 0.0006 s of time in right ascension, and about 0.006 arcsec in declination. The corresponding error in the calculated times of the phases of a solar eclipse is of the order of 1/40 s, which is much smaller than the uncertainties in predicted values of ΔT, and also much smaller than the error due to neglecting the irregularities (mountains and valleys) in the lunar limb profile. The value for ΔT (the difference between Terrestrial Dynamical Time and Universal Time) is from direct measurements during the 20th century and extrapolation into the 21st century. The value used for the Moon's mean radius is k=0.272281. These ephemerides and parameters are identical to those used in *Five Millennium Canon of Solar Eclipses: –1999 to +3000* (Espenak and Meeus 2006).

5.6 NASA Web Sites for 2010 Solar Eclipses

Two special Web sites have been set up to supplement this bulletin with additional predictions, tables, and data for the annular and total solar eclipses of 2010. Some of the data posted there include: 1) Mapping Coordinates for the Central Eclipse Path, and 2) Mapping Coordinates for the Zones of Grazing Eclipse. The URL of the NASA Web site for the 2010 annular eclipse is <http://eclipse.gsfc.nasa.gov/SEmono/ASE2010/ASE2010.html>. The URL of the NASA Web site for the 2010 total eclipse is <http://eclipse.gsfc.nasa.gov/SEmono/TSE2010/TSE2010.html>.

5.7 Predictions for Eclipse Experiments

This publication provides comprehensive information on the 2010 annular and total solar eclipses to the professional, amateur, and lay communities. Certain investigations and eclipse experiments, however, may require additional information that lies beyond the scope of this work. The authors invite the international professional community to contact them for assistance with any aspect of eclipse prediction including predictions for locations not included in this publication, or for more detailed predictions for a specific location (e.g., lunar limb profile and limb-corrected contact times for an observing site).

This service is offered for the 2010 eclipses, as well as for previous eclipses in which analysis is still in progress. To discuss individual needs and requirements, please contact Fred Espenak (fred.espenak@nasa.gov).

5.8 Algorithms, Ephemerides, and Parameters

Algorithms for the eclipse predictions were developed by Espenak primarily from the *Explanatory Supplement* (Her Majesty's Nautical Almanac Office 1974), with additional algorithms from Meeus et al. (1966), and Meeus (1989). The solar and lunar ephemerides were generated from the JPL DE200 and LE200, respectively. All eclipse calculations were made using a value for the Moon's radius of $k = 0.2722810$ for umbral contacts, and $k = 0.2725076$ (adopted IAU value) for penumbral contacts. Center of mass coordinates were used except where noted. Extrapolating from 2008 to 2010, values for ΔT of 66.0 s (annular eclipse) and 66.2 s (total eclipse) were used to convert the predictions from Terrestrial Dynamical Time to Universal Time. The international convention of presenting date and time in descending order has been used throughout the bulletin (i.e., year, month, day, hour, minute, second).

The primary source for geographic coordinates used in the local circumstances tables is *The New International Atlas* (Rand McNally 1991). Elevations for major cities were taken from *Climates of the World* (U.S. Dept. of Commerce 1972). The names and spellings of countries, cities, and other geopolitical regions are not authoritative, nor do they imply any official recognition in status. Corrections to names, geographic coordinates, and elevations are actively solicited in order to update the database for future eclipse bulletins.

AUTHOR'S NOTE

All eclipse predictions presented in this publication were generated on a Macintosh Dual 1.25 GHz PowerPC G4 computer. All calculations, diagrams, and opinions presented in this publication are those of the authors and they assume full responsibility for their accuracy.

ACRONYMS

AIDS	Acquired Immune Deficiency Syndrome
CAR	Central African Republic
CD	Compact Disk
DCW	Digital Chart of the World
DMA	Defense Mapping Agency (U.S.)
DRC	Democratic Republic of the Congo
DSLR	Digital-Single Lens Reflex
GPS	Global Positioning System
IAU	International Astronomical Union
IOTA	International Occultation Timing Association
ISO	International Organization for Standardization
ITCZ	Intertropical Convergence Zone
JNC	Jet Navigation Charts
JPL	Jet Propulsion Laboratory
NMEA	National Marine Electronics Association
ONC	Operational Navigation Charts
SASE	Self Addressed Stamped Envelope
SEML	Solar Eclipse Mailing List
SLR	Single Lens Reflex
SPCZ	South Pacific Convergence Zone
TDT	Terrestrial Dynamical Time
TP	Technical Publication
USGS	United States Geological Survey
UT	Universal Time
UV	Ultraviolet
UVA	Ultraviolet-A
WDBII	World Data Bank II

UNITS

arcmin	arc minute
arcsec	arc second
ft	foot
h	hour
Hz	Hertz
km	kilometer
m	meter
MHz	MegaHertz
min	minute
mm	millimeter
nm	nanometer
s	second

BIBLIOGRAPHY

American Conference of Governmental Industrial Hygienists Worldwide (ACGIH), 2004: *TLVs® and BEIs® Based on the Documentation of the Threshold Limit Values for Chemical Substances and Physical Agents & Biological Exposure Indices,* ACGIH, Cincinnati, Ohio, 151–158.

Bretagnon P., and G. Francou, 1988, "Planetary theories in rectangular and spherical variables: VSOP87 solution," *Astron. Astrophys.,* 202(1-2) 309–315.

Chapront-Touzé, M., and Chapront, J., 1983, "The Lunar Ephemeris ELP 2000," *Astron. Astrophys.,* **124**(1), 50–62.

Chou, B.R., 1981: Safe solar filters. *Sky & Telescope,* **62**(2), 119 pp.

Chou, B.R., 1996: Eye safety during solar eclipses—Myths and realities. In: Z. Mouradian and M. Stavinschi, eds., "Theoretical and Observational Problems Related to Solar Eclipses," *Proc. NATO Advanced Research Workshop.* Kluwer Academic Publishers, Dordrecht, Germany, 243–247.

Chou, B.R., and M.D. Krailo, 1981: Eye injuries in Canada following the total solar eclipse of 26 February 1979. *Can. J. Optom.,* **43,** 40.

Del Priore, L.V., 1999: "Eye Damage from a Solar Eclipse." In: M. Littmann, K. Willcox, and F. Espenak, *Totality, Eclipses of the Sun,* Oxford University Press, New York, 140–141.

Espenak, F., 1987: Fifty Year Canon of Solar Eclipses: 1986–2035, *NASA Ref. Pub. 1178,* NASA Goddard Space Flight Center, Greenbelt, Maryland, 278 pp.

Espenak, F., and J. Anderson, 2006: "Predictions for the Total Solar Eclipses of 2008, 2009, and 2010," *Proc. IAU Symp. 233 Solar Activity and its Magnetic Origins,* Cambridge University Press, 495–502.

Espenak, F., and J. Meeus, 2006: Five Millennium Canon of Solar Eclipses: –1999 to +3000 (2000 BCE to 3000 CE), *NASA Tech. Pub. 2006-214141,* NASA Goddard Space Flight Center, Greenbelt, Maryland, 648 pp.

Fiala, A., and M.R. Lukac, 1983: Annular Solar Eclipse of 30 May 1984, *U.S. Naval Observatory Circular No. 166,* Washington, DC, 63 pp.

Her Majesty's Nautical Almanac Office, 1974: *Explanatory Supplement to the Astronomical Ephemeris and the American Ephemeris and Nautical Almanac,* prepared jointly by the Nautical Almanac Offices of the United Kingdom and the United States of America, London, 534 pp.

Herald, D., 1983: Correcting predictions of solar eclipse contact times for the effects of lunar limb irregularities. *J. Brit. Ast. Assoc.,* **93,** 241–246.

Marsh, J.C.D., 1982: Observing the Sun in safety. *J. Brit. Ast. Assoc.,* **92,** 6.

Meeus, J., C.C. Grosjean, and W. Vanderleen, 1966: *Canon of Solar Eclipses*, Pergamon Press, New York, 779 pp.

Meeus, J., 1989: *Elements of Solar Eclipses: 1951–2200*, Willmann-Bell, Inc., Richmond, Virginia, 112 pp.

Michaelides, M., R. Rajendram, J. Marshall, and S. Keightley, 2001: Eclipse retinopathy. *Eye,* **15,** 148–151.

Morrison, L.V., 1979: Analysis of lunar occultations in the years 1943–1974, *Astr. J.,* **75,** 744.

Morrison, L.V., and G.M. Appleby, 1981: Analysis of lunar occultations–III. Systematic corrections to Watts' limb-profiles for the Moon. *Mon. Not. R. Astron. Soc.,* **196,** 1013.

Pasachoff, J.M., 2000: *Field Guide to the Stars and Planets,* 4th edition, Houghton Mifflin, Boston, Massachusetts, 578 pp.

Pasachoff, J.M., 2001: "Public Education in Developing Countries on the Occasions of Eclipses." In: A.H. Batten, Ed., *Astronomy for Developing Countries,* IAU special session at the 24th General Assembly, 101–106.

Pasachoff, J.M., and M. Covington, 1993: *Cambridge Guide to Eclipse Photography,* Cambridge University Press, Cambridge and New York, 143 pp.

Penner, R., and J.N. McNair, 1966: Eclipse blindness—Report of an epidemic in the military population of Hawaii. *Am. J. Ophthal.,* **61,** 1452–1457.

Pitts, D.G., 1993: "Ocular Effects of Radiant Energy." In: D.G. Pitts and R.N. Kleinstein, Eds., *Environmental Vision: Interactions of the Eye, Vision and the Environment,* Butterworth-Heinemann, Toronto, Canada, 151 pp.

Rand McNally, 1991: *The New International Atlas,* Chicago/New York/San Francisco, 560 pp.

Reynolds, M.D., and R.A. Sweetsir, 1995: *Observe Eclipses,* Astronomical League, Washington, DC, 92 pp.

Sherrod, P.C., 1981: *A Complete Manual of Amateur Astronomy,* Prentice-Hall, 319 pp.

Van Flandern, T.C., 1970: Some notes on the use of the Watts limb-correction charts. *Astron. J.,* **75,** 744–746.

U.S. Dept. of Commerce, 1972: *Climates of the World,* Washington, DC, 28 pp.

Watts, C.B., 1963: The marginal zone of the Moon. *Astron. Papers Amer. Ephem.,* **17,** 1–951.

Further Reading on Eclipses

Allen, D., and C. Allen, 1987: *Eclipse,* Allen and Unwin, Sydney, 123 pp.

Brewer, B., 1991: *Eclipse,* Earth View, Seattle, Washington, 104 pp.

Brunier, S., 2001: *Glorious Eclipses,* Cambridge University Press, New York, 192 pp.

Covington, M., 1988: *Astrophotography for the Amateur,* Cambridge University Press, Cambridge, 346 pp.

Duncomb, J.S., 1973: *Lunar limb profiles for solar eclipses,* U.S. Naval Observatory Circular No. 141, Washington, DC, 33 pp.

Espenak, F., and J. Meeus., 2008: Five Millennium Catalog of Solar Eclipses: –1999 to +3000 (2000 BCE to 3000 CE), *NASA Tech. Pub. 2008-214170,* NASA Goddard Space Flight Center, Greenbelt, Maryland, 270 pp.

Golub, L., and J.M. Pasachoff, 1997: *The Solar Corona,* Cambridge University Press, Cambridge, Massachusetts, 388 pp.

Golub, L., and J. Pasachoff, 2001: *Nearest Star: The Surprising Science of Our Sun,* Harvard University Press, Cambridge, Massachusetts, 286 pp.

Harrington, P.S., 1997: *Eclipse!,* John Wiley and Sons, New York, 280 pp.

Harris, J., and R. Talcott, 1994: *Chasing the Shadow: An Observer's Guide to Solar Eclipses,* Kalmbach Publishing Company, Waukesha, Wisconsin, 160 pp.

Littmann, M., F. Espenak, and K. Willcox,, 2008: *Totality, Eclipses of the Sun,* Oxford University Press, New York, 341 pp.

Mitchell, S.A., 1923: *Eclipses of the Sun,* Columbia University Press, New York, 425 pp.

Meeus, J., 1998: *Astronomical Algorithms,* Willmann-Bell, Inc., Richmond, Virginia, 477 pp.

Meeus, J., 1982: *Astronomical Formulae for Calculators,* Willmann-Bell, Inc., Richmond, Virginia, 201 pp.

Mobberley, M., 2007: *Total Solar Eclipses and How to Observe Them,* Astronomers' Observing Guides, Springer, New York, 202 pp.

Mucke, H., and Meeus, J., 1983: *Canon of Solar Eclipses: –2003 to +2526,* Astronomisches Büro, Vienna, Austria, 908 pp.

North, G., 1991: *Advanced Amateur Astronomy,* Edinburgh University Press, Edinburgh, Scotland, 441 pp.

Ottewell, G., 1991: *The Understanding of Eclipses,* Astronomical Workshop, Greenville, South Carolina, 96 pp.

Pasachoff, J.M., 2004: *The Complete Idiot's Guide to the Sun,* Alpha Books, Indianapolis, Indiana, 360 pp.

Pasachoff, J. M., 2007: "Observing solar eclipses in the developing world. In: Astronomy in the Developing World, *Proc. IAU Special Session 5,* J.B. Hearnshaw and P. Martinez, eds., Cambridge University Press, New York, 265–268.

Pasachoff, J.M., and B.O. Nelson, 1987: Timing of the 1984 total solar eclipse and the size of the Sun. *Sol. Phys.,* **108,** 191–194.

Steel, D., 2001: *Eclipse: The Celestial Phenomenon That Changed the Course of History,* Joseph Henry Press, Washington, DC, 492 pp.

Stephenson, F.R., 1997: *Historical Eclipses and Earth's Rotation*, Cambridge University Press, New York, 573 pp.

Todd, M.L., 1900: *Total Eclipses of the Sun*, Little, Brown, and Co., Boston, Massachusetts, 273 pp.

Von Oppolzer, T.R., 1962: *Canon of Eclipses*, Dover Publications, New York, 376 pp.

Zirker, J.B., 1995: *Total Eclipses of the Sun*, Princeton University Press, Princeton, New Jersey, 228 pp.

Further Reading on Eye Safety

Chou, B.R., 1998: Solar filter safety. *Sky & Telescope*, **95**(2), 119.

Pasachoff, J.M., 1998: "Public education and solar eclipses." In: L. Gouguenheim, D. McNally, and J.R. Percy, Eds., New Trends in Astronomy Teaching, *IAU Colloquium 162* (London), Astronomical Society of the Pacific Conference Series, 202–204.

Further Reading on Meteorology

Griffiths, J.F., Ed., 1972: *World Survey of Climatology, Vol. 10, Climates of Africa*, Elsevier Pub. Co., New York, 604 pp.

National Climatic Data Center, 1996: *International Station Meteorological Climate Summary; Vol. 4.0* (CD-ROM), NCDC, Asheville, North Carolina.

Schwerdtfeger, W., Ed., 1976: *World Survey of Climatology, Vol. 12, Climates of Central and South America,* Elsevier Publishing Company, New York, 532 pp.

Wallen, C.C., Ed., 1977: *World Survey of Climatology, Vol. 6, Climates of Central and Southern Europe*, Elsevier Publishing Company, New York, 258 pp.

Warren, S.G., C.J. Hahn, J. London, R.M. Chervin, and R.L. Jenne, 1986: Global Distribution of Total Cloud Cover and Cloud Type Amounts Over Land. *NCAR Tech. Note NCAR/TN-273+STR* and *DOE Tech. Rept. No. DOE/ER/60085-H1*, U.S. Department of Energy, Carbon Dioxide Research Division, Washington, DC, (NTIS number DE87-006903), 228 pp.

REQUEST FORM FOR NASA ECLIPSE BULLETINS

NASA eclipse bulletins contain detailed predictions, maps and meteorology for future central solar eclipses of interest. Published as part of NASA's Technical Publication (TP) series, the bulletins are prepared in cooperation with the Working Group on Eclipses of the International Astronomical Union and are provided as a public service to both the professional and lay communities, including educators and the media. In order to allow a reasonable lead time for planning purposes, subsequent bulletins will be published 12 to 24 months before each event. Comments, suggestions and corrections are solicited to improve the content and layout in subsequent editions of this publication series.

Single copies of the bulletins are available at no cost and may be ordered by sending a 9 x 12 inch SASE (self addressed stamped envelope) with sufficient postage for each bulletin (12 oz. or 340 g). Use stamps only since cash or checks cannot be accepted. Requests within the U. S. may use the Postal Service's Priority Mail. Please print either the eclipse date (year & month) or NASA publication number in the lower left corner of the SASE and return with this completed form to either of the authors. Requests from outside the U. S. and Canada may use ten international postal coupons to cover postage. Exceptions to the postage requirements will be made for international requests where political or economic restraints prevent the transfer of funds to other countries. Professional researchers and scientists are exempt from the SASE requirements provided the request comes on their official or institutional stationary.

Permission is freely granted to reproduce any portion of this NASA Reference Publication. All uses and/or publication of this material should be accompanied by an appropriate acknowledgment of the source.

Request for: NASA TP-2008-214171—Annular and Total Solar Eclipses of 2010

Name of Organization: _____
(in English, if necessary): _____
Name of Contact Person: _____
Address: _____
_____
City/State/ZIP: _____
Country: _____
E-mail: _____

Type of organization: ___ University/College   ___ Observatory   ___ Library
(check all that apply) ___ Planetarium   ___ Publication   ___ Media
                       ___ Professional   ___ Amateur   ___ Individual

Size of Organization: _____ (Number of Members)

* *

Return Requests   Fred Espenak   or   Jay Anderson
and Comments to: NASA/GSFC   Royal Astronomical Society of Canada
                     Code 693   189 Kingsway Ave.
                     Greenbelt, MD 20771   Winnipeg, MB,
                     USA   CANADA R3M 0G4

                     E-mail: fred.espenak@nasa.gov   E-mail: jander@cc.umanitoba.ca

2008 Oct

www.ingramcontent.com/pod-product-compliance
Lightning Source LLC
Chambersburg PA
CBHW081835170526
45167CB00007B/2813